AF414143

PHYTOTOXICANTS

PHYTOTOXICANTS

Shatrughna Sukdeo Patole

Head, Department of Zoology
S.G. Patil Arts, Science and Commerce College
Sakri
Maharashtra

MJP Publishers

MJP PUBLISHERS

© Publishers, 2024   47, Nallathambi Street
All rights reserved   Triplicane
Chennai 600 005

This book has been published in good faith that the work of the author is original. All efforts have been taken to make the material error-free. However, the author and publisher disclaim responsibility for any inadvertent errors.

To

My Ancestors

PREFACE

India is one of the world's 12 regions having the largest biodiversity. It has 16 agro-climatic zones and 45,000 plant species, of which 15,000 to 20,000 possess proven biocidal properties. Plants have the capacity to synthesize a wide variety of biochemicals. The toxic constituents, viz., alkaloid, glycosides, saponin, terpenoids, essential oils, flavonoids, etc., present in plants are secondary metabolites and have an insignificant role in the primary physiological processes of plants. They are reportedly defensive and are known to disrupt behaviour and physiology of pests in various ways and even prove toxic to different developmental stages.

This book is not intended to be comprehensive but to put across the basic principles of the subject, Phytotoxicants, as briefly and lucidly as possible. The aim is to enthuse the reader with this active and exciting area of research and to lay a solid foundation on which further research or study of its various facets may be based. I have freely consulted various scientific journals in the preparation of this book, and have tried to make it as comprehensive and up to date as possible.

Though every care has been taken by me and the publisher to bring out an error-free book, it is quite likely that some errors might have found their way into the book. But I hope that these are not very significant. However, suggestions to improve the book and pointing out of errors and mistakes, if any, will gratefully be appreciated.

I am deeply indebted to my teacher Dr. R. T. Mahajan, Reader, Post-graduate Department of Zoology, M. J. College, Jalgaon (M.S.), for his valuable suggestion and guidance. I would also like to thank Dr. G. B. Kabra, Reader, Entomology Department, Mahatma Phule Krishi Vidhyapith, Rahuri (M.S.), for checking the references and the taxonomical aspects and for proof-reading. I also thank Dr. P. D. Deore, Principal, and Prof. D. L. Torwane, Dean, Faculty of Mental and Moral Sciences, North Maharashtra University, Jalgaon, for their constant inspiration.

I am indebted to my wife, Sau. Ratnamala Patole, daughter, Sayali, and son, Palash, who have showered their affection and encouraged me during my long hours of work in preparing the manuscript.

Last but not least, I am grateful to the Managing Editor and the editorial team of MJP Publishers, Chennai, for their valuable cooperation and interest in publishing this book.

It is earnestly hoped that the present book will prove to be useful to beginners in the subject, students of agrticulture/plant sciences, pharmacy, chemistry and other allied subjects, teachers and scientists.

Shatrughna Sukdeo Patole

CONTENTS

1

CHEMISTRY OF TOXICANTS

1.1 INTRODUCTION

Toxicants or toxic substances are defined by the National Institute for Occupational Safety and Health as those which "demonstrate the potential to induce cancer, produce long-term disease or bodily injury, affect health adversely; produce acute discomfort; or endanger the life of an organism through exposure via the respiratory tract, skin, eyes, mouth or other routes." When such toxic ingredients are extracted from plant sources they are termed as phytotoxicants.

Plants serve as the most important and formidable pest control agents since the time of our forefathers. Plants are one of the foremost organisms to invade the planet and hence humans had to live with these plant species which are seen everywhere. Plants by their diverse nature continue to provide food to man, animals and microbes and be useful to prevent colonization of insects and other herbivores. The leaves, seeds, fruits, stem, roots and bark of plants are useful in checking the problems of pests in the field. Drug powder, oils and extracts of different parts of the plants are used traditionally in pest control.

Even before we had synthetic pesticides, plants like neem, chrysanthemum, *Acorus*, *Vitex*, *Adhatoda*, onion, chilli, turmeric, tobacco, *Derris*, ginger, chinaberry, etc. were used by ancient humans in their health care, animal welfare and crop production. This feature is continued traditionally (Narayansamy, 2006).

1.2 HISTORICAL PERSPECTIVES

We understand from the available literature that during the Vedic era, there was the practice of abundant use of plants in health care, and plant protection as evidenced by the care, love and nurturing of plants and trees by the Vedic masters. Many of the Rishis, Siddars and Saints were known to have preserved the knowledge of secrets of personal achievements but had not made them available to the common man or to the experimental practitioners perhaps due to the lack of proper understanding of the concepts of healing. The plant species, which were commonly used in the pest control activities during the Vedic period, are mentioned below.

Common name	Botanical/Scientific name
Ajasringi	*Odina pinnata*
Asoka	*Saraca indica*
Aswatha	*Ficus religiosa*
Arka	*Calotropis gigantea*
Kakadasingi	*Gynandropsis pentaphylla*
Kuverakshi	*Bignonia suareolens*
Maharaksha	*Euphorbia tirucalli*
Neem	*Azadirachta indica*
Nyagrotha	*Ficus indica*
Palasa	*Butea frondosa*
Stapuspa	*Anethum sowa*
Sikhandi (Juha)	*Embelia ribes*
Vidanga	*Jasminum auriculatum*

In the use of plants in pest control, the various modes of action of the plants on insects are as follows:

i. Killing the pest species.

ii. Acting as antifeedant against the pests.

iii. Protecting the plants themselves from the pest degradation.

iv. Inhibiting reproductive, alimentary, neurotic and other physiological processes of the insect body system.

v. Effecting juvenilitic changes in the body morphology of insect stages.

vi. Acting as contact and stomach poison.

All these would show the traditional wisdom of the use of plants in insect pest control.

1.3 DISCOVERY OF INSECTICIDES IN PLANTS

Long before the development of synthetic organic insecticides, natural substances derived from plants were successfully employed in pest control. Today, about 2000 plants containing insecticidal properties are known (Perkow, 1968), but only a negligible small number of these plants are cultivated, their products being used commercially on a large scale. Scruples about the undesirable side-effects of modern synthetic insecticides, (such as, high and acute toxicity, long degradation periods of some chlorinated hydrocarbon insecticides, their concentration in food chains, the suspected dangers of chronic poisoning through the continuous intake of small quantities and the undesirable extension of their power to destroy both useful insects and pests) raise the question of how far it is possible to substitute harmless natural insecticides for harmful ones. In brief, they have been used for some time. Domestically used sprays have contained as major constituents vegetable substances, in particular, the pyrethrin of certain chrysanthemums, combined with other agents, the synergists, and these are harmless to man and domestic animals.

The beginning of the "Agricultural Revolution" (Cipolla, 1975), ten to fifteen thousand years ago, brought about a considerable change.

The protection of food stores was of considerable importance to those who gathered in harvests, and who harvested large amounts of wild plants. But noxious insects less easily attack seeds and fruits of wild plants than that of contemporary crops. Thus, for the first time, the planned cultivation of wild plants and the consequent development of crop plants with a higher susceptibility to pests was realized ten thousand years ago (Schwanitz, 1967), and this stimulated thought and action of man in purposefully fighting against the continuous losses of the crops in the fields and in store. Naturally, the "Agricultural Revolution", and the selection of crop plants with higher yields, led to an increase in settlement density (Cipolla, 1975), with the resulting need to increase storage and protect their contents. There are written records of experiences in this respect in Egypt, Greece, China and other civilized areas, three or four thousand years ago (Mayer, 1959) and the methods of protecting crops and stores are described, these being mainly limited to the employment of inorganic materials like lime, arsenic, wood ashes, etc. besides organic mixtures.

From 400 BC up to the 16th century, we have independent reports from Asia and Europe of the development of information on the therapeutic influences of medicinal plants in books on medicinal plants (Pharmacopoeias). These statements and recipe books contained contemporary and locally restricted information on the plants and parts of plants that could be used to combat pests (Bock, 1577). However local knowledge was not disseminated. Moreover the succeeding "Age of Discovery" did not, in general, materially contribute to the dissemination of natural drugs, even though a number of cultivated crops, as, for example, the potato and tobacco, were relatively early known to the explorers of the New World or imported into Europe.

Nevertheless, now and then, navigators of the "Age of Discovery" reported that, during long voyages, they were not infected by lice that plague sailors, because of the use of insect powder. Up to the 19th century, however, all this information on the effects of plants on insect had no further influence. Boccone's first report (1697) on the herbs, *Chrysanthemum marschallii* and *Chrysanthemum roseum,* which occur in Transcaucasia, and which yielded an insect powder, was almost forgotten, until at the beginning of the 19th century the Armenian, Jumtikoff, observed that certain Chrysanthemum blossoms were used as an insecticide, and, from 1828, his son successfully exported this "Caucasian insect powder" (Casida, 1973).

The Austrian, Anna Rosauer, discovered its insecticidal effects in 1840 from the dead flies lying around dried Chrysanthemum from Dalmatia, and this led to the cultivation of the plant to be used as "Dalmatian insect powder". The rapid development of scientific knowledge in the 19th century, discoveries of bacteriology, hygiene and other medical disciplines, knowledge of the transmission of disease by insects, and the initiation of plant protection, all favoured the commercial employment of insecticidal plants. In addition to Pyrethrum (various Chrysanthemum species), quassia extracts and other natural materials from plants were introduced to combat insects.

Alexander von Humboldt, also, in his journeys to South America (1799–1804), did not describe in detail the use of roots as fish poisons (barbasco, a collective name for roots, which contained not only fish poisons, but also insecticides), which like the East Asian derris roots (tuba) were employed by native people occasionally as an insecticide (Brehm, 1914).

In the second half of the 19th century, suggestions made by Oxley in 1848 to employ derris roots as a fish poison, as an ingredient of arrow poison and as an insecticide were not heeded (Muller, 1954). It was not until the 20th century that this advice was again taken up, although Chinese gardeners in the Singapore area had employed successfully the diluted milky juice of fresh derris roots a hundred years earlier (Kempski, 1940). Only in the final years of the 19th century were large quantities of vegetable insecticides (mainly pyrethrum) employed in the United States in increasing quantities in agriculture, for the protection of stores and within the home. The planned cultivation in their original homes, and in later cultivated areas, in Dalmatia, Japan, and East Africa, was delayed greatly by the transport difficulties caused by the two world wars. Apart from the reduction in production caused by war, and despite the introduction of synthetic insecticides like DDT, the employment of natural materials has continued to increase rapidly and constantly. On the one hand the explanation of the structure of the main plant insecticides, the pyrethrins, begun by Staudinger and Rudzicka in 1924, and of the rotenoids of derris roots by many research groups (Muller, 1954), initiated intensive research in this area, and on the other hand, people had not completely renounced the use of these harmless expedients in households and to protect stores. In addition, they soon began to employ the strong synergistic effects of certain natural materials in combination with the newly discovered synthetic compounds. Some of the important insecticidal plant species and their uses are shown in Table 1.1.

TABLE 1.1   USES OF SOME INSECTICIDAL PLANTS

Plant family	Plant species	Activity
Aesculaceae	*Aesculus california*	Insecticide
Annonaceae	*Annona reticulate* *Annona squamosa*	Rotenone-like effect (seed)
Apocynaceae	*Haplophytum cimidium*	Alkaloid with insecticidal effect (leaves)
Boraginaceae	*Heliotropum peruvianum* *Tournefortia hirsutissima*	Insecticide
Cannaceae	*Canna* species	House fumigants (leaves and stems)
Celastraceae	*Tripterium wilfordii*	Insecticide (roots)
Chenopodiaceae	*Anabasis aphylla*	Insecticide (leaves and stem)
Clusiaceae	*Mammaea americana*	Insecticide (seed and bark)
Cochlospermaceae	*Cochlospermum gossypium*	Synergist in plant gum
Asteraceae	*Chrysanthemum cinerariae-folium*, etc. *Heliopsis scabra*	Pyrethrin, insecticide in blossoms and pyrethrin type material
Cucurbitaceae	*Cucurbita pepo*	Repellent and insecticide (leaves and seeds)
Euphorbiaceae	*Croton tiglium* *Ricinus communis*	Insecticide and fish poison Synergists (seeds)
Flacourtaceae	*Ryania speciosa*	Insecticide (wood)
Fagaceae	*Castanea dentata*	Repellent
Labiatae	*Ocimum basilicum* *Salvia officialis*	Contact poison and repellent (leaves, roots and seeds)

(*Contd.*)

TABLE 1.1 (CONTINUED)

Plant family	Plant species	Activity
Leguminosae	*Haematoxylon campechianum*	Repellents, contact poison (seed and roots)
	Millettia	Rotenoid (bark)
	Pachyrhizus erosus	Insecticide, piscicide (seed)
	Tephrosia virgiana	Contact poison
	Derris elliptica	Rotenoid (roots)
	Derris malaccensis *Lonchocarpus* species	Rotenoids (roots), insecticides and fish poison
Liliaceae	*Amianthium muscaetoxicum* *Melanthium virginicum* *Schoenocaulon officinale*	Insecticide (leaves and seeds)
	Veratum album *Veratum viride*	Insecticide (roots)
Meliaceae	*Melia azedarach*	Insecticide and repellent (leaves and berries)
Myrtaceae	*Pimenta racemosa*	Insecticide, attractant (berries) and repellent
Pedaliaceae	*Sesamum indicum*	Synergists (seeds)
Ranunculaceae	*Delphinium consolida*	Insecticide (alkaloids in seed)
Rutaceae	*Phellodendron amurense*	Insecticides (fruit)
	Xanthoxylum clavaherculis	Insecticides (bark)
Sapindaceae	*Sapindus marginatus*	Repellent (berries)
Simaroubaceae	*Ailanthus* species	Insecticides (wood and bark)
Solanaceae	*Duboisia hopwoodi* *Micandra physalodes* *Nicotiana tabacum* *Nicotiana rusticum* *Nicotiana glauca*	Insecticides (alkaloids in leaves)
Stemonaceae	*Stemona tuberosa*	Insecticides (roots)
Umbelliferae	*Carum carvi* *Coniummaculatum* *Coriandrum sativum* *Pimpinella anisum*	Insecticides and repellents (seeds)
Vitaceae	*Parthenocissus quinquefolia*	Insecticides

SOURCE: AFTER NEGHERBON, 1959

The failure of the new compounds resulting from the development of resistance, after a relatively short period of use, stimulated the production of mixed preparations of natural and synthetic synergists (a drug/chemical that interact with another drug/chemical to produce increased activity, the effect being greater than the sum of the effects of the two drugs, given separately) which at least to a limited extent, helped to ward off the development of resistance in certain noxious insects.

1.4 ORDER OF IMPORTANCE

In India there are about 700 poisonous species of plants belonging to over 90 families of flowering plants. The more important of these in their order of importance are Ranunculaceae, Euphorbiaceae, Leguminosae, Solanaceae, Asteraceae, Apocynaceae, Asclepiadaceae, Ericaceae, Liliaceae, Gramineae, Araceae, Anacardiaceae, Thymeleaceae, Rosaceae and Rubiaceae. As in the case of medicinal plants, it does not mean that other families do not contain important poisonous plants. In fact there are several other families which have a few very potent species belonging to them.

1.5 BOTANICAL INSECTICIDES

The insecticides from different plant species as well as families mentioned in Table 1.1, have varied effects (Negherbon, 1959). These include the following.

1. The essential oils extracted from Umbelliferae plants like caraway, coriander, etc. showed pronounced repellent activity. It shows synergistic effect when used with insecticidal alkaloids found in hemlock, and nicotine from *Nicotiana* species.

2. Alkaloids like nicotine and nor-nicotine, act as food, contact and respiratory poisons and are found in varying concentrations and relationships in various plant families like Solanaceae and Chenopodiaceae.

3. Rotenone-containing plants (*Derris elliptica*, *Lonchocarpus* species), and the wood and bark of *Ailanthus* and *Quassia* are effective insecticides.

4. Pyrethrin obtained from *Chrysanthemum* species act as highly contact insecticides and also exist in highly active vegetable synergists, which when used singly have hardly any insecticide effect, but when used as additives, increase the effect of other insecticides.

Thus, in principle, earlier workers find the same gradations in effect and possibilities of use in plant insecticides as in synthetic organic materials (Table 1.2).

Practically all plant-based insecticides are completely degraded in a short period, and without leaving any residue. Even nicotine, which is quite comparable with the very highly toxic phosphoric acid esters, quickly becomes harmless. Enrichment in food chains, as in the case of chlorinated hydrocarbon insecticides, is not known for natural substances with insecticidal effect.

TABLE 1.2   COMPARISON BETWEEN SYNTHETIC AND NATURAL COMPOUNDS

Synthetic compounds	Natural compounds
The first generation	
Phosphoric acid esters with very acute toxicity and broad spectrum	Nicotine and other alkaloids
Chlorinated hydrocarbons with less acute toxicity and broad spectrum	Pyrethrin and rotenone
The second generation	
Materials with lower toxicity with a selective effect	Ryania, quassia and others
The third generation	
Materials with juvenile hormone effect, sexual aromatic substances and other biologically active substances able to specifically destroy particular pests	Phytoecdysone and juvenile hormone and specific repellents from plants

For the future, it will be most interesting to know how far substances with the juvenile hormone effect can be isolated from plants (Interference with larval development). In contrast to the isolation of juvenile hormones from insects, obtaining them from plant materials has the advantage that, using a larger amount of initial raw material, one can proceed more easily and more cheaply, though, in fact, even when using vegetable material, the discovery of micro-quantities of the effective principle still involves many great difficulties and much expense.

Up to now, it is not known how far the effects of certain woods, like sandalwood, earlier described as repellent, and of which clothes, chests and cupboards were made, acting as a protection against moths, are due to the juvenile hormone factor. Other plants combine an insecticidal and repellent effect, an example being basil, which contain a mosquito and fly poison, while at the same time, in many Mediterranean countries, where it is grown in pots on balconies, it apparently produces essential oils which quite effectively repel mosquitoes. Some plant leaves, like those for instance, used to envelop certain cheeses in France, influence not only the fermentation and flavour of the cheese, but prevent the cheese fly from laying eggs. House and cookery books of the last century advised the treatment of homemade small cheese with some plant leaves.

Many biological methods useful at home cannot in practice be adopted in a wider context, let alone in an industrial programme. For example, a simple domestic device is employed in Greece against nematodes. A watery garlic extract is poured upon plants grown in pots; the tiny soil nematode worms escaping from the wet soil are collected by hand. But this cannot be introduced naturally and successfully into contemporary agriculture.

Some main vegetable insecticides are shown in Table 1.3. It includes:

* The alkaloids nicotine and anabasine have, because of their extreme toxicity when used as a tobacco infusion, etc., declined to a mere token use.

* Synergists derived from pepper and sesame seeds, etc. are of less importance than chemically related, cheaper synthetic products like piperonyl butoxide.

* Insecticides with a certain limited and selective effect from Veratrum, Ryania, Tripterygium and Quassia are still used in

specific instance, but they are not substances of international significance (Martin, 1967; Perkow, 1968).

* Pyrethrum and derris preparations are by far, the most commonly employed vegetable insecticides, and in their countries of origin, they are an important factor in the economy (Casida, 1973; Shepard, 1951).

1.6   TOXIC CONSTITUENTS OF PLANTS

The poisonous properties of a plant, as has been pointed out, are due to the presence of certain toxic constituents. Our knowledge of the chemical constituents of plants has advanced rapidly in recent years, and the constituents responsible for the specific physiological action of the plant have in many cases been isolated, purified, and identified as definite chemical compounds. The pharmacological action of many of these has been studied by modern methods of assay and we are now in possession of a more exact knowledge of the action of the drugs containing these constituents. A good deal of work, however, has yet to be done in connection with the isolation and elucidation of the chemical nature of many other toxic or physiologically active constituents known to be present in various plants. Quite a large number of plants are known or suspected to be toxic, few are investigated but still a large number remains uninvestigated and this is particularly the case with Indian flora.

The major toxic constituents of plants are listed in the following sections.

1.6.1 Vegetable Bases

An important group of compounds constitute the nitrogenous vegetable bases, which include amines, purines and alkaloids.

Amines   Amines are simpler natural bases, mainly derived from amino acids, the building material of all proteins. Some amines give a foetid odour to the plants in which they occur and the poisonous character to some mushrooms. The foetid odour of *Paederia foetida* Linn. is due to the presence of indole. Several bacteria are known to produce physiologically active amines when they act on

TABLE 1.3   THE MAIN VEGETABLE INSECTICIDES

Substances	Chemical groups	Originating plant species with part content	Effect
Nicotine	Alkaloid	*Nicotiana tabacum, Nicotiana rustica* (leaves 5–14%) and a further 9 alkaloids	Food, contact and respiratory poison. Fatal dose for man is 60 mg, while 4 mg dose produces serious illness.
Anabasine	Alkaloid	*Anabasis aphylla* (leaves 1–2.6%)	Nicotinelike effect.
Piperidline	Alkaloid	*Piper nigrum* (seeds and other plants)	Synergist, practically innocuous to vertebrates.
Veratrine	Alkaloid	*Schoenocaulon officinale* (seeds 2–4%), *Veratrum album*	Selective contact and food insecticides (10 time effective as (root) DDT), and also poisonous to man.
Ryanodine	Alkaloid	*Ryania speciosa* (wood 0.16–0.20%)	Oral food poison (stomach poison) Selective effect. Contact poison. Very low toxicity to vertebrates.
Wilfordine	Alkaloid (a mixture of 5 alkaloids)	*Tripterygium wilfordii* (root)	Selective food poison, for pests of stores; low toxicity to vertebrates.
Quassin, Neoquassin, Picrasmin	Diterpenoids Lactones	*Quassia amara, Picrasama excelsa* (wood)	Selective action. No toxic effect on vertebrates.
Sesamin	The crystalline fraction of sesame oil (0.25%)	*Sesamum indicum* (seeds)	Synergist. Very low toxicity to vertebrates.
Rotenone (ellipton, sumatrol, malaccol, deguelin, α-toxicarol)	Rotenoids	*Derris (Deguelia elliptica)* (root)	Contact poison, food poison (Fatal dose for man 2000–3000 mg a person)
Pyrethrin I, Pyrethrin II, Cinerin I, Cinerin II, Jasmolin I, Jasmolin II.	Pyrethrins	*Chrysanthemum cinerariaefolium, Chrysanthemum rosesum Chrysanthemum carneum* (flowers, 0.7–3%)	Contact poison, practically non-toxic to man and domestic animals.

the proteins of various foodstuff, producing poisoning. Such amines are however, comparatively rare among the higher plants.

Purines or methylxanthins　Another class of nitrogenous compounds occur as the active principles of some tropical plants, such as tea (*Camellia sinensis* Linn. Kuntez), coffee (*Coffea arabica* Linn.), cocoa (*Theobroma cacao* Linn.), kola (*Cola acuminata* Schott and Endl. and *Cola vera* K. Schum), Guarana (*Paullinia cupana* H. B. and K.), etc. They are confined only to a few genera. The important representatives of these purines having some physiological action are caffeine, theobromine, theophylline, etc. The purines may be considered as the derivatives of a parent substance named purin, which on hydrogenation is converted into purine, $C_5H_4N_4$.

Alkaloids　Alkaloids form the most important group of vegetable bases. They are complex heterocyclic nitrogenous compounds having a basic nature and are mostly tertiary amines. They are found in plants in combination with various organic acids as salts, which make them more or less soluble in water. Some of the alkaloids are nontoxic, but as a class they are characterized by a profound physiological action, and in many cases by their intensely poisonous nature. As a rule, they have a bitter taste which is frequently, for the plants containing them, a sufficient protection against being eaten by livestock. No alkaloids have been found in algae, not many in vascular cryptogams and Gymnosperms, but more in monocotyledons while the vast majority of them have been reported from dicotyledons. Rich alkaloid-containing families are: Ranunculaceae, Papaveraceae, Leguminosae, Rubiaceae, Apocynaceae, Solanaceae, Liliaceae, etc. Examples of typically toxic alkaloids are aconitine from aconite roots (species of *Aconitum*), morphine from poppy capsules (*Papaver somniferum* Linn.), emetine from ipecacuanha roots (*Psychotria ipecacuanha* Stokes), strychnine from nux-vomica seeds (*Strychnos nux-vomica* Linn.), nicotine from tobacco leaves (species of *Nicotiana*), coniine from the hemlock (*Conium maculatum* Linn.) and curarine from curare (*Strychnos toxifera* Schomb ex Benth).

1.6.2 Glucosides

These are widely distributed in plants, than alkaloids. They are defined as compounds which when split up with the help of acids or enzymes yield a sugar or some closely allied carbohydrate and one or more of other products (usually phenols, aldehydes, alcohols or acids), and therefore also known as "aglucones". Many do not contain nitrogen. Some modern authors designate these compounds as "glycosides" and restrict the name "glucosides" to only those in which the sugar compound is glucose. The term glucoside has, therefore, been used throughout this work. The glucoside can basically be divided into two types—nontoxic and toxic. The toxic glucoside may either directly be toxic to man or animals or give rise to toxic components on hydrolysis.

Of the latter, the most important class comes under the term "Cyanogenetic" glucosides. These glucosides, although by themselves more or less harmless, give rise to the most toxic acid, hydrocyanic acid. Examples of these cyanogenetic glucosides are amygdalin found in bitter almond, phaseolunatin found in flax, prunasin found in wild cherry, sambunigrin found in the elder and so on. Of particular economic and toxicological interest is the occurrence of cyanogenetic glucosides in a number of grasses (Gramineae). Other glucosides which may be said to yield somewhat harmful components on hydrolysis are sinigrin found in the black-mustard seeds, sinalbin found in the white mustard seeds, etc. These yield essential oils containing various sulphur compounds which are irritant in their action, whereas some glucosides, viz., digitoxin found in *Digitalis*, cerebrin found in *Cerebra*, strophanthin found in *Strophanthes*, theretin from *Theretia*, Paristyphnin found in *Paris* and antiarin from *Antiaris*, etc. may be said to have direct toxic action on animals.

1.6.3 Saponins

The saponins occur in about 400 species belonging to 50 different families. The saponins produce soapy foam on being shaken with water, thus their name. They possess a bitter, acrid taste, and in the form of a dry powder are very irritating to the nose. They are particularly toxic to cold-blooded animals, such as fishes, frogs, insects, etc. Fishes are killed in such high dilution as 1 : 200,000. In warm-blooded animals, they often produce gastro-intestinal irritation, vomiting, and diarrhoea when taken by mouth. They produce haemolysis when they come in contact with blood. The more poisonous saponins are sometimes known as "sapotoxins". On hydrolysis they yield different sugars, generally hexoses and pentoses, and another component known as sapogenin, which is often physiologically active.

1.6.4 Bitter Principles

These are neutral and nonglucosidal in nature and possess a bitter taste, are found in a number of plants and are of common occurrence in the wild members of the Cucurbitaceae family. Preparations of several plants containing these are used in therapeutics to increase the appetite, and some have purgative properties.

1.6.5 Toxic Proteins

The toxic proteins or "toxalbumins" have been observed in plants of the families Leguminosae, Euphorbiaceae, etc. Examples of toxalbumins are abrin, crotin, ricin and curcin, etc. These are essentially blood poisons, and are characterized by their property of agglutinating and precipitating the red blood corpuscles.

1.6.6 Fixed Oils

These are compounds of glycerol with different kinds of fatty acids containing sterols and other substances dissolved in them. They are greasy liquids occurring quite commonly in seeds of plants. When heated, they decompose giving off acrid acrolein vapours. They are insoluble in water but soluble in organic solvents. They generally have laxative properties and have drastic purgative action.

1.6.7 Essential Oils

The essential or volatile oils are odorous principles, which are generally responsible for the odour of plants in which they occur. They usually occur as such in plants, but in some cases they are found in a state of combination with glucosides from which they may be liberated by the interaction of enzymes, as is known to occur in some members of the family Cruciferae. The essential oils differ from the fixed oils by being volatile in steam. They are generally mixtures of different chemical compounds which may include hydrocarbons known as terpenes and sesquiterpenes, open-chain alcohols and aldehydes, aromatic alcohols of the camphor series and their ketones, sesquiterpene alcohols, phenols and their derivatives, esters of different alcohols, and sulphur compounds.

From the nature of their components, it may be expected that essential oils would act as antiseptics, disinfectants, insect repellents, and in some cases as insecticides. They possess a sharp burning taste and locally have an irritant action which is especially marked upon the mucous membranes. Large doses given internally produce a violent irritation of the entire gastro-intestinal tract with consequent pain, vomiting and diarrhoea. The hyperemia may spread to the peritoneum and to all neighbouring parts, among other, in the female, to genital organs, where the congestion may find expression in haemorrhage and abortion. This especially happens in the case of certain oils, such as those of Juniper, savin, rue, pennyroyal and parsley. Some of the essential oils, e.g. that of wormwood or absinthe, nutmeg, etc. act directly on the central nervous system. The activity of the oils as nervous poisons varies greatly. In most cases the central nervous system is first stimulated and later depressed. In the case of absinthe, however, there is only a marked excitement resulting in convulsions.

The plants bearing essential oils are distributed widely in the vegetable kingdom while certain families, such as Labiatae, Piperaceae, Rutaceae, Umbelliferae, Myrtaceae, Lauraceae and Coniferae are especially rich in such plants. Among the plants containing essential oils with toxic constituents may be mentioned species of the genera *Artemisia, Ruta, Mentha, Petroselinum, Chenopodium, Myristica, Eucalyptus, Gaultheria, Juniperus, Pinus, Eupatorium, Anemone, Ranunculus, Caltha, Prunus, Allium, Brassica* and other crucifers, *Piper, Ferula*, etc.

1.6.8 Resins

The main constituents of many resins are esters known as resin esters, complex acids known as resin acids, and substances of unknown constitution known as resenes. Some are found to contain phenolic substances with strongly irritant properties, while others contain bitter substances and have a strong purgative action.

1.6.9 Organic Acids

From the poisoning point of view, the most significant organic acid is oxalic acid, a protoplasmic poison, occurring in a large number of plants in the form of oxalates of calcium, sodium, and potassium. As a rule, oxalates occur in plants in too small quantities to produce poisoning. Another organic acid is formic acid, an irritant substance, has also been found to occur in a few plants, e.g. in the stinging nettles of the Urticaceae family.

1.6.10 Tannins

The term tannin is used to denote a group of phenol derivatives distinguished by their giving a bluish or greenish colour with ferric salts. They are non-nitrogenous; some are glucosides. They have an astringent action and occur in many plants, especially in leaves, barks and in pathological formations (nutgalls).

REFERENCES

Casida, J.E. (ed.). (1973). *Pyrethrum: The Natural Insecticides.* Academic Press, New York and London.

Cipolla, C.M. (1975). *Economic History of the World Population,* revised (edn.). Penguin, Harmondsworth.

Frank Beye. (1978). Insecticides from Vegetable Kingdom. *Plant Research and Development.* Institute for Wissen Schaftliche, Germany. 7: 13–31.

Hieronymus, Bock. (1577). Kreutterbuch (Herbal). Rihel, Strasbourg.

Kempski, K.E. (1940). The Insecticidal Plants: Pyrethrum, Derris, Barbasco. *The Hamburg Tropical Series*, 45.

Martin, H. (1967). *The Scientific Principles of Crop Protection*. Translation of 5th revised edition. Verlag Chemie, Weinheim.

Muller, P. (1954). The chemistry of insecticides, its development and its present state. In: *Experientia*. 10: 91–131.

Narayansamy, P. (2006). Traditional wisdom of plants in pest controls in agriculture. In: *Medicinal Plants Traditional Knowledge*. Pravin Chandra Trivedi (ed.). I.K. International Publishing House Pvt. Ltd. New Delhi, India. pp. 159–184.

Negherbon, W.O. (1959). *Handbook of Toxicology. Vol III: Insecticides*. W.B. Saunders, Philadelphia and London.

Perkow, W. (1968). *The Insecticides*, 2nd (edn.). A. Hüthig Verlag, Heidelberg.

Schwanitz, F. (1967). *The Evolution of Cultivated Plants*. Munich Basel, Vienna, BLV,

Shepard, H.H. (1951). *The Chemistry and Action of Insecticides*. McGraw Book Co., New York, Toronto and London.

2

ECO-FRIENDLY BOTANICALS AS TOMORROW'S PESTICIDES: AN OVERVIEW

2.1   INTRODUCTION

Generally toxic substances produced by plants yield insecticides that can be used for insect eradication. Apart from these toxic substances, the plant world has an abundant source of biochemicals that could be tapped as eco-friendly pesticides. These effective constituents present in the plant represent the secondary metabolites that have only an insignificant role in primary physiological processes in the plant that synthesizes them (Sharma, 1982; Ashalata *et al.*, 1999). These metabolites include terpenoids, alkaloids, polyacetylenes, flavonoids, unusual amino acids, glycosides, phenols, tannins, etc. Such natural products repel the approaching insect, deter feeding and oviposition, disrupt the behaviour and physiology of insects in various ways and even prove lethal to the different developmental stages of many pests. Some important, eco-friendly botanicals found worldwide were used to protect agricultural crops and stored grains from insects and other pests (Jacobson and Crosby, 1971).

The pesticides which are less harmful to the environment as well as to human health and have benign impact on the environment during the various stages of development, manufacture, use, packing, distribution, consumption, disposal and recycling may be called eco-friendly products and will be labelled with the Eco mark label. A large number of safe biodegradable pesticides are of botanical origin and have numerous advantages in comparison to synthetic pesticides. The main advantage is their quick bio-degradability and thus total safety to environment (Sharma, 1982; Roy, 1996). A broad classification of chemicals obtained from plants with their effects on insects and their possible role in pest management is illustrated in Figure 2.1.

Among natural products, some are not directly beneficial for the growth and development of the organism. These compounds are usually regarded as parts of the plant's defence against plant-feeding pests and other herbivores (Rosenthal and Janzen, 1979; Arora and Dhaliwal, 1994).

Secondary metabolites of plants can be sub-divided into kairomones and allomones. The kairomones include general or specific attractants and pheromonal analogues or precursors, which serve to attract the insect and can be used for population monitoring or selective eradication in combination with insecticides and chemosterilants. The allomones cover a broad spectrum of activities such as repellence, antifeedant, oviposition deterrence, hormonal, antihormonal, chitin inhibition, etc. These may affect insect behaviour, physiology, development and reproduction and can be used in a number of ways for pest management. Systemic antifeedants and antimetabolites or allelochemicals constitute yet another category affecting insect nutrition and causing general antibiosis, which can also be utilized in judicious combinations with other principles and measures for effective pest control. Finally, plant products such as lignin, bark, gum, gelatins, saw dust, cellulose and bacterial fermentation products can be used as biodegradable, non-pollutant and slow-release carriers of pest control agents (Sharma, 1982).

Potential of Plant Products for Pest Control

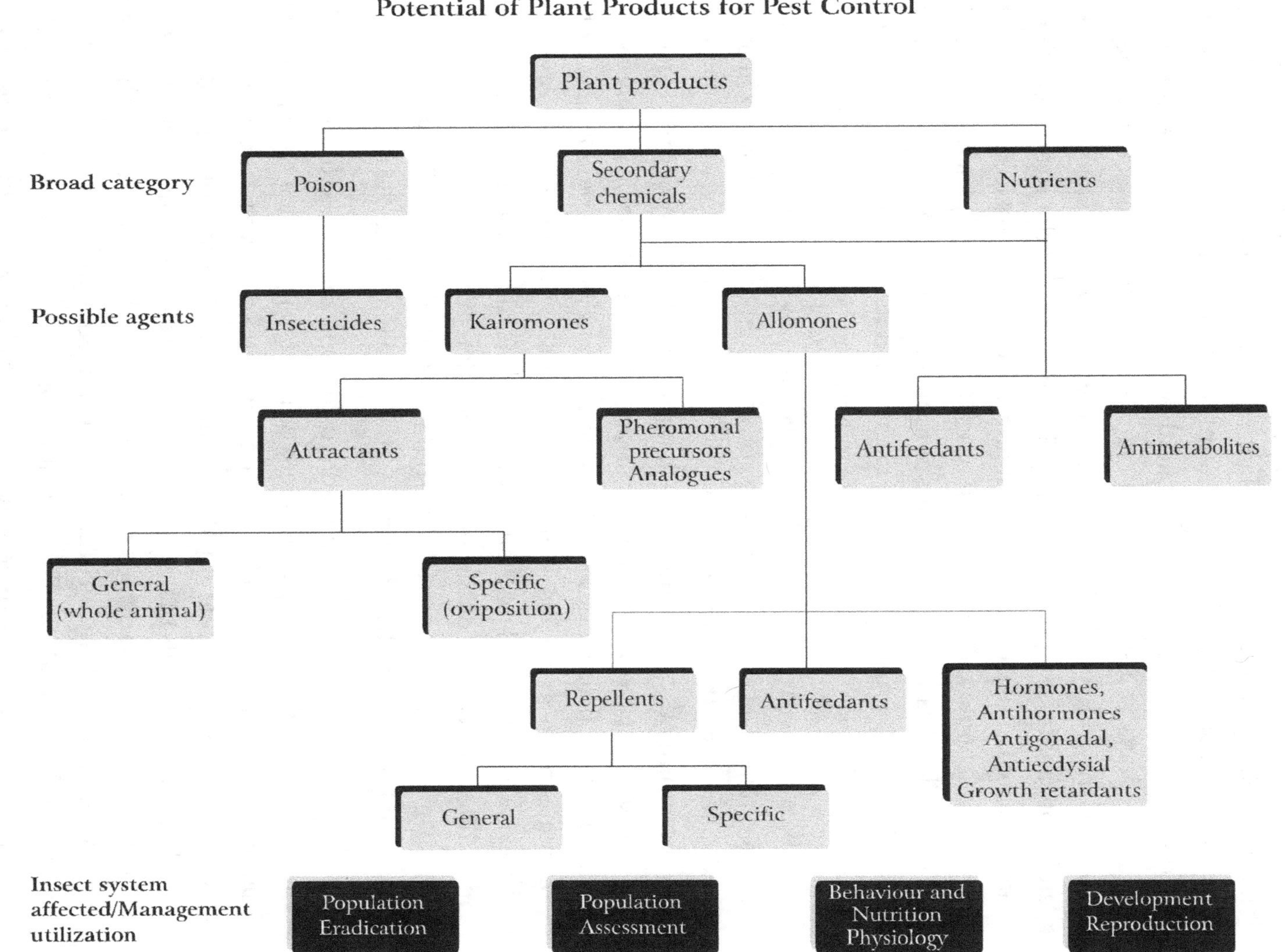

FIGURE 2.1   A BROAD CLASSIFICATION OF BOTANICALS WITH THEIR EFFECTS ON INSECTS AND POSSIBLE ROLE IN PEST MANAGEMENT (*SOURCE:* SHARMA, 1982)

Plant-based insecticides are of great interest to many, because they are natural insecticides. Historically, plant material has been in use longer than any other group, with the possible exception of sulphur. Tobacco, pyrethrum, derris, hellebore, quassia and terpenoids were some of the more important plant products in use before the organized search for insecticides began. Some of the widely used insecticides come from plants. The flower, leaves, and roots are finely ground and used in this form, or the toxic ingredients are extracted and used alone or in mixtures with other toxicants. There are five natural botanicals that are of interest to gardeners in general, especially to the organic gardener. These are pyrethrum, rotenone, sabadilla, ryania and nicotine. Botanical insecticide use reached its maximum in the United States in 1966 and has declined steadily since then (Manju Yadav, 2004).

In recent years, there has been an increased interest in natural plant-derived materials as alternatives to conventional broad-spectrum chemical pesticides or synthetic insecticides. The natural indigenous plant materials are economical, perennial, non-hazardous, target-specific, innately biodegradable, eco-friendly, non-persistent and easily available in comparison with chemical insecticides (Prakash and Rao, 1990; Saxena *et al.*, 1992; Saxena and Tiwary, 1994; Isman *et al.*, 2002). The important aspect of these natural material is their wide range of effects against insect pests including repellence, as antifeedants, oviposition deterrence, toxicity, sterility, and as growth inhibitors (Khaire *et al.*, 1992; Ascher, 1993; Schmutterer and Rembold, 1995; Tripathy *et al.*, 2001; Ignacimuthu, 2004).

Here an attempt has been made to compile available data particularly on botanical pesticides or insecticides of plant origin which shall provide a database to conduct research on related aspects of insect pest control. This chapter contains information of some important international plant-based eco-friendly botanicals that have been reported earlier as insecticidal, insect repellent, antifeedant, oviposition-deterrent, feed deterrent, etc. These are anabasine, azadirachtin, nicotine, pyrethrum, quassin, rotenone, ryanodine, wilfordine and sabadilla. The advantage of such indigenous sources is the formulation of an eco-friendly natural pesticide, which may prove their utility in controlling the pests generally in the agricultural and public health sectors. This attempt will also boost organic farming and rural development.

2.2   SOME BOTANICAL INSECTICIDES

1.   Anabasine (Alkaloid)

Status	None
IUPAC	
CAS	3-(2*S*)-2-piperidinylpyridine
Reg. No.	494-52-0
Formula	$C_{10}H_{14}N_2$
Activity	Botanical insecticide
Structure	

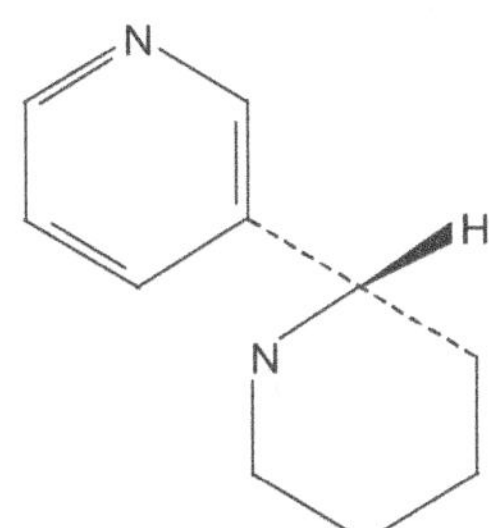

Plant Source	*Anabasis aphylla* L. (Leaves), Chenopodiaceae
Importance	There is no ISO common name for this substance; the name "anabasine" has been used in the literature but has no official status. It is a food, contact and respiratory poison. The fatal dose for humans is 60 mg, while 4 mg produces serious illness. It is obtained from the leaves and stem of the plant *Anabasis aphylla* L. (Chenopodiaceae). It is an alkaloid also found along with nicotine and nor-nicotine from tobacco leaf. Anabasine has some insecticidal properties, but its relative abundance and relative toxicity are far less as compared to nicotine.

2. Azadirachtin

Status   None

IUPAC

CAS   Dimethyl (2a*R*, 3*S*, 4*S*, 4a*R*, 5*S*, 7a*S*, 8*S*, 10*R*, 10a*S*,10b*R*)-10-(acetyloxy) octahydro-3,5-dihydroxy-4- methyl-8-[[(2*E*)-2-methyl-1-oxo-2-butenyl]oxy]-4-[1a*R*,2*S*,3a*S*,6a*S*,7*S*,7a*S*)- 3a,6a,7,7a-tetrahydro-6a-hydroxy-7a-methyl-2,7-methanofuro[2,3-*b*]oxireno[*e*]oxepin-1a (2*H*)-yl]-1*H*,7*H*-naphtho[1,8-*bc*:4,4a-*c*′] difuran-5,10a(8*H*)-dicarboxylate

Reg. No.   11141-17-6

Formula $C_{35}H_{44}O_{16}$

Activity   Botanical insecticide

Structure

Plant Source   *Azadirachta indica* A. Juss (Seed kernel), Meliaceae

Importance   There is no ISO common name for this substance; the name "azadirachtin" has been used in the literature but has no official status. It is safe, and 100 per cent insecticidal, and nontoxic to humans. All parts of the neem tree possess insecticidal activity but seed kernel is the most active. Neem products exhibit almost every conceivable type of activity against insects. It has been reported to be antifeedant, attractant, repellent, nematicide, insecticide and growth disrupter. Therefore, commercialization of neem-based pesticides is expanding at a rapid rate. More than three dozen products are marketed in India. Neem has assumed the status of an international tree which is evident from the fact that it has been a subject of discussion at six global conferences, viz; Rottach-Egern, Germany (1980), Rauischolzhausen, Germany (1983), Nairobi, Kenya (1986), Bangalore, India (1993), Queensland, Australia (1996) and Vancouver, Canada (1999). It was also discussed in the International Plant Protection Congress at Beijing, China during 2004.

3. Rotenone (Rotenoids)

Status	ISO 765
IUPAC	(2*R*, 6a*S*, 12a*S*)-1, 2, 6, 6a,12,12a-hexahydro-2-isopropenyl-8,9-dimethoxychromeno[3,4-*b*]furo [2,3-*h*]chromen-6-one
CAS	(2*R*,6a*S*,12a*S*)-1,2,12,12a-tetrahydro-8,9-dimethoxy-2-(1-ethylethenyl)[1] benzopyrano[3,4-*b*] furo [2,3-*h*][1] benzopyran-6(6a*H*)-one
Reg. No	83-79-4
Formula	$C_{23}H_{22}O_6$
Activity	Botanical insecticide

Structure

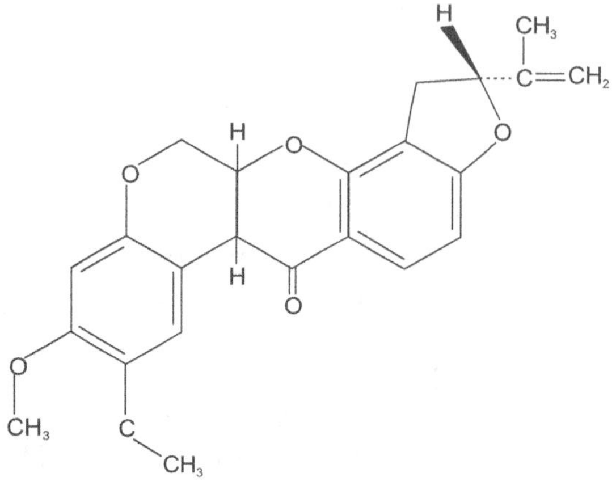

Plant Source   *Derris elliptica* Roxb. (Root), Leguminosae

Importance   This substance is considered by the International Organization for Standardization to not require a common name. The name "derris" is approved by the Japanese Ministry of Agriculture, Forestry and Fisheries. It acts as a contact and food poison, the fatal dose for man being 2000–3000 mg per person. It is highly poisonous to a large number of the insect pests except mammals which are little affected by it. It is the most important insecticide procured from the roots of certain leguminous plants. Since long, plants containing rotenone have been made use of as a fish poison in South America (at least since 1649). But its use as an insecticide against leaf-eating caterpillar was first realized in 1848. Rotenone easily decomposes when it is left in the sunlight and air and for that matter, it can safely be used even on eatables. It puts under control both piercing, sucking and chewing insects. It is a selective non-systemic insecticide with some acricidal properties. The other naturally occurring rotenoids are elliptone, sumatrol, malaccol, toxicarol, deguelin and tephrosine, etc.

Rotenone kills the insect by attacking the nervous system, causing paralysis. It acts as a contact poison. It gives better performance in high relative humidity, which retards evaporation when applied in liquid form. Rotenone is probably second in its use as botanical insecticide, the first being pyrethrum.

4. Nicotine (Alkaloid)

Status ISO 765

IUPAC (S)-3-(1-methylpyrrolidin-2-yl)pyridine

CAS 3-[(2S)-1-methyl-2-pyrrolidinyl]pyridine

Reg. No 54-11-5

Formula $C_{10}H_{14}N_2$

Activity Botanical insecticide

Structure

Plant Source *Nicotiana tabacum* L. (Leaves), Solanaceae

Importance When nicotine is used as a salt, its identity should be stated, for example nicotine sulphate. It acts as a food, contact and respiratory poison. The fatal dose for man is 60 mg, while 4 mg produces serious illness. Its use as an insecticide dates back to the 17th century, long before chemists isolated and characterized nicotine as the major toxic principle of tobacco. Commercial preparation of nicotine sulphate was put into market by 1910 and has been a popular insecticide ever since, by its triple action insecticidal property acting as stomach, contact and fumigant poison. One of the advantages of the insecticidal use of nicotine is its reported high margin of safety, easier handlability and much less toxicity to warm-blooded animals. Because of the high volatile nature of nicotine, insecticidal nicotine preparation leaves no appreciable residue on treated plants. The alkaloid is believed to disappear in relatively short time usually in minutes thus leaving no hazardous material on the marketable produce. Nicotine is variably and widely used against piercing or sucking insects which do not get destroyed by stomach poisons. It is generally used as a killing agent for aphids and other soft-bodied insects. Dusts are toxic to humans hence are not available for garden use.

5. Pyrethrum/Pyrethrins

Chrysanthemum is an almost cosmopolitan genus comprising about 300 species of herbs and undershrubs of which only a few yield the commercial insecticides. The term pyrethrum is commonly applied to the dried flower heads of any of the three species, e.g. *Chrysanthemum cinerariaefolium*, *Chrysanthemum coccineum* (Syn *C. roseum*) and *Chrysanthemum marshallii*. In India *C. cinerariaefolium* and *C. coccineum* are reported to be cultivated. Pyrethrum is one of the safest insecticides. It has very low mammalian toxicity (LD_{50} 1500 mg/kg body weight) and it is rapidly metabolized if accidently swallowed. For insects, it acts as a contact poison with a rapid "knock-down" effect on a wide range of insect species. It affects the nervous system and results in muscular excitation, convulsion, paralysis, instantaneous knock-down effect and death. Therefore, it is desirable for the control of pests that affect public health, stored food and household. The active ingredients of pyrethrum degrade rapidly on exposure to sunlight.

Pyrethrum is indeed an ideal insecticide, and it is therefore not surprising that a bulk of pyrethrum is used in or around homes. About 70% of the world production finds its way into fly-sprays and insecticidal aerosols, 20% is used in mosquito repellent coils and the balance in other formulations. However, it is always used with some synergist, the most popular being piperonyl butoxide prepared from oil of sassafras, a natural product.

The six effective components in the flowers of the Asteraceae species, *C. cinerariaefolium, C. roseum* and *C. marshallii* are Pyrethrin I and II, Cinerin I and II and Jasmolin I and II. In Pyrethrin I, the alcohol pyrethrolone is esterified with chrysanthemum-monocarboxylic acid, and in Pyrethrin II with chrysanthemum-dicarboxylic acid. In the case of cinerins, the same acids and esters are esterified with the alcohol cinerolene, and in the Jasmolins, with jasmolone. All these six terpene esters are viscous oils, not soluble in water. They are very freely oxidized because they are unsaturated. The diene side-chains are very susceptible to oxidation. Their international status, chemical formulae and structures, etc. are as follows.

i.   Pyrethrin I (Pyrethrins)

Status   ISO 765

Reg. No   121-21-1

Formula $C_{21}H_{28}O_3$

Activity   Botanical insecticides

Structure

(Z)–(S)–alcohol (1R)–trans-acid

Importance   This substance is considered by the International Organization for Standardization not to require a common name; it is one of the components of pyrethrins. It is a contact poison, practically non-toxic to man and domestic animals.

ii.   Pyrethrin II (Pyrethrins)

Status   ISO 765

IUPAC   (*Z*)-(*S*)-2-methyl-4-oxo-3-(penta-2,4-dienyl) cyclopent-2-enyl(*E*)-(1*R*,3*R*)-3-(2-thoxycarbonylprop-1-enyl)-2, 2-dimethylcyclopropanecarboxylate or(*Z*)-(*S*)-2-methyl-4-oxo-3-(penta-2,4-dienyl)cyclopent-2-enyl (*E*)-(1*R*)-*trans*-3-(2-methoxycarbonylprop-1-enyl)-2,2-dimethylcyclopropanecarboxylate) or (*Z*)-(*S*)-2-methyl-4-oxo-3-(penta-2,4-dienyl) cyclopent-2-enyl pyrethrate

CAS   (1*S*)-2-methyl-4-oxo-3-(2*Z*)-2,4-pentadienyl-2-cyclopenten-1-yl (1*R*,3*R*)-3-[(1*E*)-3-methoxy-2-methyl-3-oxo-1-propenyl]-2,2-dimethyl-cyclopropanecarboxylate

Reg. No   121-29-9

Formula $C_{22}H_{28}O_5$

Activity   Botanical insecticides

Structure

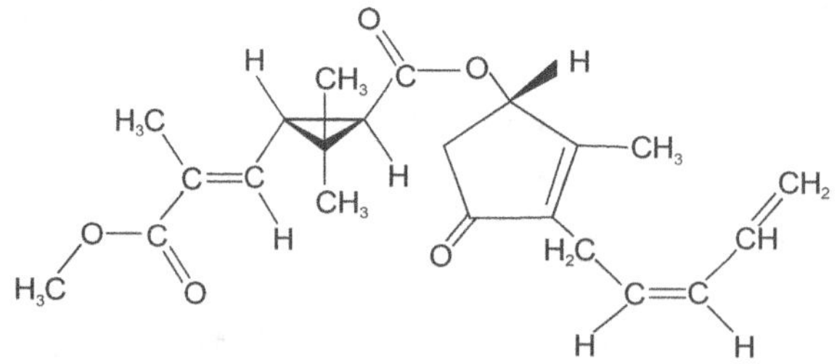

(*Z*)-(*S*)-alcohol (E)-(1R)-*trans*-acid

Importance	This substance is considered by the International Organization for Standardization not to require a common name; it is one of the components of pyrethrins.

iii. Cinerin I (Pyrethrins)

Status	ISO 765
IUPAC	(*Z*)-(*S*)-3-(but-2-enyl)-2-methyl-4-oxocyclopent-2-enyl (1*R*,3*R*)-2,2-dimethyl-3-(2-methylprop-1-enyl)cyclopropanecarboxylateor(*Z*)-(*S*)-3-(but-2-enyl)-2-methyl-4-oxocyclopent-2-enyl (1*R*)-*trans*-2, 2-dimethyl-3-(2-methylprop-1-enyl)cyclopropane-carboxylate or (*Z*)-(*S*)-3-(but-2-enyl)-2-methyl-4-oxocyclopent-2-enyl(+)-*trans*-chrysanthemate
CAS	(1*S*)-3-(2*Z*)-2-butenyl-2-methyl-4-oxo-2-cyclopenten-1-yl (1*R*,3*R*)-2,2-dimethyl-3-(2-methyl-1-propenyl) cyclopropane-carboxylate
Reg. No	25402-06-6
Formula	$C_{20}H_{28}O_3$
Activity	Botanical insecticides
Structure	

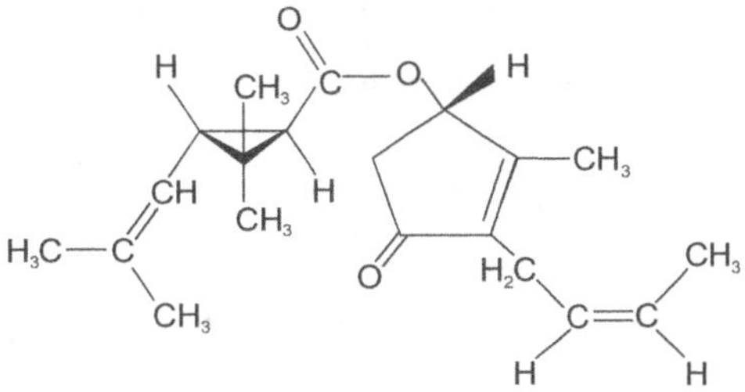

(*Z*)-(*S*)-alcohol (1R)-*trans*-acid

Importance	This substance is considered by the International Organization for standardization not to require a common name. It is one of the components of

pyrethrins. The mixture of cinerin I and cinerin II is known as cinerin, which is considered by the British Standards Institution not to require a common name.

iv. Cinerin II (Pyrethrins)

Status ISO 765

IUPAC (Z)-(S)-3-(but-2-enyl)-2-methy l-4-oxocyclopent-2-enyl (E)- (1R,3R)-3-(2-methoxycarbonylprop-1-enyl)-2,2-dimethyl-cyclopropanecarboxylate or (Z)-(S)-3-(but-2-enyl)-2-methyl-4-oxocyclopent-2-enyl (E)-(1R)-*trans*-3-(2-methoxycarbonylprop-1-enyl)-2,2-dimethylcyclopropanecarboxylate or (Z)-(S)-3-(but-2-enyl) -2-methyl-4-oxocyclopent-2-enyl pyrethrate

CAS (1S)-3-(2Z)-2-butenyl-2-methyl-4-oxo-2-cyclopenten -1-yl (1R,3R)-3-[(1E)-3-methoxy-2-methyl-3-oxo-1-propenyl]-2,2-dimethylcyclopropanecarboxylate

Reg. No 121-20-0

Formula $C_{21}H_{28}O_5$

Activity Botanical insecticides

Importance This substance is considered by the International Organization for Standardization not to require a common name. It is one of the components of pyrethrins.

Structure

(Z)-(S)-alcohol (E)-(1R)-*trans*-acid

v. Jasmolin I (Pyrethrins)

Status	None
IUPAC	(*Z*)-(*S*)-2-methyl-4-oxo-3-(pent-2-enyl)cyclopent-2-enyl (1*R*,3*R*)-2,2-dimethyl-3-(2-methylprop-1-enyl)cyclo- propanecarboxylate or (*Z*)-(*S*)-2-methyl-4-oxo-3-(pent-2-nyl)cyclopent-2-enyl (1*R*)-*trans*-2,2-dimethyl-3-(2-methylprop-1-enyl) cyclopropanecarboxylate or (*Z*)-(*S*)-2-methyl-4-oxo-3-(pent-2-enyl)cyclopent-2-enyl (+)-*trans*-chrysanthemate
CAS	(1*S*)-2-methyl-4-oxo-3-(2*Z*)-2-pentenyl-2-cyclopenten-1-yl 1*R*,3*R*)-2,2-dimethyl-3-(2-methyl-1-propenyl)cyclopropanecarboxylate
Reg. No.	4466-14-2
Formula	$C_{21}H_{30}O_3$
Activity	Botanical insecticides
Structure	

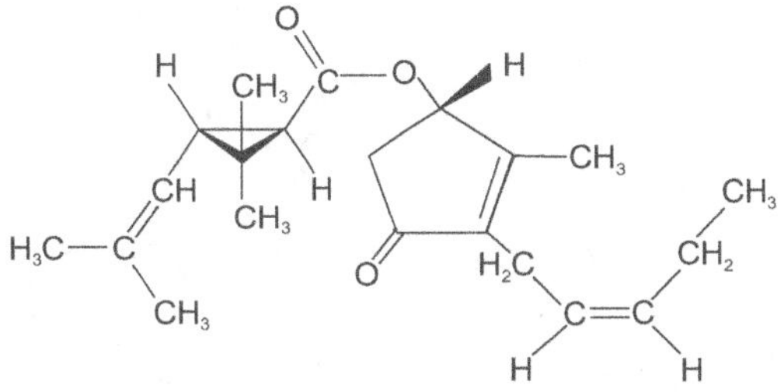

(Z)-(S)-alcohol (1R)-*trans*-acid

Importance	There is no ISO common name for this substance; the name "Jasmolin I" has been used in the literature but has no official status. It is one of the components of pyrethrins.

vi.   Jasmolin II (Pyrethrins)

Status	None
IUPAC	(Z)-(S)-2-methyl-4-oxo-3-(pent-2-enyl)cyclopent-2-enyl (E)-(1R,3R)-3-(2-methoxycarbonylprop-1-enyl)-2,2-dimethylcyclo- propanecarboxylate or (Z)-(S)-2-methyl-4-oxo-3-(pent-2-enyl) cyclopent-2-enyl (E)-(1R)-*trans*-3-(2-methoxy-carbonylprop-1-enyl)-2,2-dimethylcyclopropanecarboxylate (Z)-(S)-2-methyl-4-oxo-3-(pent-2-enyl)cyclopent-2-enyl pyrethrate
CAS	(1S)-2-methyl-4-oxo-3-(2Z)-2-pentenyl-2-cyclopenten-1-yl (1R,3R)-3-[(1E)-3-methoxy-2-methyl-3-oxo-1-propenyl]-2,2-dimethyl-cyclopropanecarboxylate
Reg. No.	1172-63-0
Formula	$C_{22}H_{30}O_5$
Activity	Botanical insecticides
Structure	

(Z)-(S)-alcohol (E)-(1R)-*trans*-acid

Importance	There is no ISO common name for this substance; the name "Jasmolin II" has been used in the literature but has no official status. It is one of the components of pyrethrins.

6. Quassin (Diterpenoids)

Status	None
Reg. No	—
Formula	—
Activity	Botanical insecticides
Importance	There is no ISO common name for this extract of the plant *Quassia amara* L. (Simaroubaceae); the name "quassia" has been used in the literature but has no official status. It has selective action and has no toxic effect on vertebrates. It appears very likely that even before the turn of the 19th century, aqueous extracts were in use against insects. Due to its water-soluble ingredients, quassia sprays are used against sap feeders when taken up by the roots. Quassia works systematically and is transported into the leaves where it acts as a stomach poison. The plant acts as a contact, systemic and stomach poison. It has insecticidal as well as larvicidal properties. The entire plant has insecticidal properties, but the roots, leaves and bark contain Quassin to a smaller degree. In India, farmers use *Picrasma excelsa*, which is either closely related to or identical with *Aeschrion excelsa*. Beneficial insects like ladybirds and honeybees are not killed by quassia spray.

7. Ryanodine (Alkaloid)

Status	None
Reg. No.	8047-13-0
Formula	$C_{25}H_{35}NO_9$
Activity	Botanical insecticides

Importance	There is no ISO common name for this extract of the root and stem of the plant *Ryania speciosa* Vahl. (Flacourtiaceae), a shrub grown in Trinidad. The name "ryania" has been used in the literature but has no official status and its chemical structure is still not determined. One of the main alkaloidal ingredients is ryanodine [15662-33-6]. It acts as a contact and stomach poison, and shows selective action against insect pests. It is a safe insecticide, quite safe for human and domestic animals. It has an oral LD_{50} of approximately 750 mg/kg. It is a slow-acting insecticide, requiring as long as 24 hours to kill. Insects exposed to ryania usually stop their feeding almost immediately, therefore, ryania may act as a good antifeedant for caterpillars. The efficacy of Rynodine for holding lepidopterous larvae in check is beneficial. It is soluble in water, methyl alcohol and most organic solvents except petroleum oils. Its residue appears to be less toxic to man than nicotine and rotenone. It is more stable and possesses a longer residual action.

8. Sabadilla/Veratrine (Alkaloid)

Status	None
Reg. No.	8051-02-3
Formula	$C_{32} H_{49} O_9 N$
Activity	Botanical insecticides
Importance	There is no ISO common name for this extract of the plant *Schoenocaulon officinale* Grey (Liliaceae); the names "sabadilla" and "cevadilla" have been used in the literature but have no official status. Sabadilla ranks among the oldest insecticides. It is procured from the seeds of the plant *S. officinale* Grey.

The insecticidal action of sabadilla is due to cevadine ($C_{32}H_{49}O_9N$) and veratridine ($C_{36}H_{51}O_{11}N$), the two known active alkaloids. Their chemical structures have not been established. Both are contact and stomach poisons to insects. Its oral LD_{50} is approximately 5000 mg/kg, making it the least toxic to warm-blooded animals. But it is irritating to humans and may cause eye irritation and violent sneezing in some sensitive individuals. It deteriorates rapidly in sunlight and can be used safely on food crops with no waiting interval required. Sabadilla is registered for most commonly grown vegetables and will control caterpillars, grasshoppers, beetles, leafhoppers, thrips, chinch bugs, stink bugs, etc.

9. Wilfordine (Alkaloid)

Status	None
Reg. No.	—
Formula	—
Activity	Botanical insecticides
Importance	There is no ISO common name for this extract of the plant *Tripterygium wilfordii* Hooke F. (Celastraceae); the name "Wilfordine" has been used in the literature but has no official status. It acts as a selective food poison for pests within stores, and has low toxicity to vertebrates. *Tripterygium wilfordii* has been employed widely in China as a garden insecticide known as *lei-kung-teng*. Aqueous extracts have a direct effect on the heart, smooth and striated muscles rather than primarily on the nervous system. Although some species of chewing insects are repelled by bark, the stomach-poison effect is quite selective. When

lepidopteran larvae were fed on treated plant, they were violently affected and often shrunk to less than 1/3rd of their original size. However, flaccid paralysis was fairly rapid, and the insects frequently could recover when removed to untreated food. Although the insecticide has been claimed to be safe for man and domestic animals, it has little contact action against aphids. The root bark extracts show insecticidal property. It contains at least five insecticidal ester alkaloids.

2.3.   CONCLUSION

Botanicals, because of their origin, are biodegradable and they do not leave residues or by-products that contaminate the environment. This safety feature is very important because it has an impact to a large extent on the cost of development and registration of a new pesticide product. The research and development cost of botanical pesticides from discovery to marketing is much less compared to chemical pesticides. The naturally occurring pesticides appear to have a prominent role in the development of future commercial pesticides not only for agricultural crop productivity but also for the safety of the environment and public health. We, in India, are favourably placed with regards to the availability of different pesticidal plants for the manufacture of effective pesticidal products for use in agriculture. Some foreign firms are keen to produce botanicals, but Indian firms should be encouraged in these directions by imposing some relaxation in registration of such pesticidal products.

Globally, more than 2000 plant species from different families have been documented to possess insecticidal properties. In India some of flowering plants have been reported for their insecticidal activities. Some of the families like Ranunculaceae, Euphorbiaceae, Leguminosae, Solanaceae, Asteraceae, Apocynaceae, Asclepiadaceae, Ericaceae, Liliaceae, Poaceae, Araceae, Anacardiaceae, Thymeleaceae, Rosaceae and Rubiaceae possess very potent insecticidal plants. Considering the wealth of flora and plant-based resources available in our country, the work on detecting insecticidal principles among plants has received little attention. The rising cost of petroleum-based insecticides, their non-biodegradability and lipophilicity cause them to be stored in the cells (biomagnification) and to interfere with the oxidation and energy production. Problems of environmental hazards and increasing resistance by insect pests will definitely make a big way for exploring possibilities and gaining greater awareness for the botanical insecticides. These insecticides definitely become prospective of tomorrow's eco-friendly pesticides.

2.4 SUGGESTIONS

* The plant diversity with insecticidal properties needs to be explored and harnessed for their use as botanical pesticides.

* Research on the combined effects of different botanicals should be encouraged.

* Awareness on the use and advantages of botanicals must be developed and motivated among users especially farmers.

* The NGOs should come forward and guide farmers in areas regarding importance of botanicals and also in developing simple formulations.

* The botanicals should possess more deterrence rather than repellence.

* There should be information centers in universities and colleges and also a good link among researchers, industries and botanical pesticide users, i.e., farmers.

* Private entrepreneurs should be encouraged.

* Botanical pesticides should be good alternatives to chemical pesticides and it should be eco-friendly, economical and target-specific and biodegradable.

REFERENCES

Arora, R. and Dhaliwal, G.S. (1994). Botanical pesticides in insect pest management. In: Dhaliwal, G.S. and Kansal, B.D. (eds.). *Management of Agricultural Pollution in India*. Commonwealth Publishers, New Delhi, India. pp. 213–245.

Ascher, K.R.S. (1993). Nonconventional insecticidal effects of pesticides available from the neem tree, *Azadirachta indica*. *Arch. Insect. Biochem. Physiology.* 22: 433–449.

Ashalata D'Rozario, Subir Bera and Dipak Mukherji. (1999). *A Hand Book of Ethnobotany*. Kalyani Publishers, Ludhiana.

Ignacimuthu, S. (2004). Green pesticides for insect pest management. *Current Science.* Vol.86 (4).

Isman, M.B., Arnason, J.T. and Towers, G.H.N. (2002). Chemistry and biological activity of ingredients of other species in Meliaceae. In: *The Neem Tree*, Schmutterer, H. (ed.). Neem Foundation, Mumbai. 652–666.

Jacobson Martin and Crosby, D.G. (1971). *Naturally Occurring Insecticides*. Marcel Dekker, Inc. New York.

Khaire, V.M., Kachare, B.V. and Mote, U.N. (1992). Efficacy of different vegetable oils as grain protectants against the pulse beetle, *Callosobruchus chinensis* L. in increasing storability of pigeon pea. *J. Stored Prod. Res.* 28: 153–156.

Manju Yadav. (2004). *Applied Entomology*. Discovery Publishing House, New Delhi.

Prakash, A. and Rao, J. (1990). Leaves of begunia pulse grain protectant. *Indian J. Entomol.* 51: 192–195.

Rosenthal and Janzen, (1979). Using natural pesticides: Current and future prospectives. A report for the plant protection improvement programme in Bostwana, Zambia and Tanzania.

Roy, N.K. (1996). *Ecofriendly pesticides: Its impact in agriculture. Recent Advances in Indian Entomology*. Lal, O.P. (ed.). APC Publication Pvt. Ltd., New Delhi. pp. 38–42.

Saxena, R.C. and Tiwary, A. (1994). Repellent and feeding deterrent activity of *Spaheranthus indicus* L. against *Tribolium castaneum* (Herbst). *Bio. Science Research Bulletin.* 9(1–2): 57–59.

Saxena, R.C., Dixit, P. and Harsar, V. (1992). Insecticidal action *Lantena camera* against *Callosobruchus chinensis* (Coleoptera: Bruchidae). *J. Stored Prod. Res.* 28: 279–281.

Schmutterer, H. and Rembold. (1995). Reproduction. In: *The Neem tree Azadirachta indica A Juss. and other meliceous plants*: Weinheim, Federal Republic of Germany. pp. 195–204.

Sharma, R.N. (1982). Development and utilization of plant products for insect control- A comprehensive approach. *Cultivation and utilization of medicinal plants.* Atwal, C.K. and Kapur, B.M. (eds.) RRL, CSIR, Jammu-Tawi. pp. 657–667.

Tripathy, M.K., Sahoo, P., Das, B.C. and Mohanty, S. (2001). Efficacy of botanical oils, plant powders and extracts against *Callosobruchus chinensis* L. attacking black gram (Cv. T9). *Legume Research.* 24(2): 82–86.

3

A SURVEY OF PROMISING INSECTICIDAL PLANT SPECIES

3.1 INTRODUCTION

Plants are a rich source of bioactive organic chemicals. It has been estimated that there are about 2,50,000 to 7,50,000 plant species in the world. Only 5 to 15% of these have been surveyed for biologically active compounds indicating that there is large scope. The total number of plant chemicals may exceed 400,000. Of these, 10,000 are secondary metabolites whose major role in the plant is reportedly defence. Over the last 50 years, 2400 plant species from 189 plant families and genera were reported to contain toxic principles and are pesticidal plants (Singh, 2000). Plants are known to produce a diverse range of secondary metabolites such as terpenoids, alkaloids, polyacetylenes, flavonoids, unusual amino acids, sugars, etc. The structures of more than 600 alkaloids, 3000 terpenoids, several thousands of phenylpropanoids, 1000 flavonoids, 500 quinones, 650 polyacetylenes and 4000 amino acids have already been elucidated (Metcalf and Metcalf, 1992).

The importance of the conservation of environment quality and the preservation of the extent of and natural equilibrium in an ecosystem are now well-acknowledged. On the other hand there are incontrovertible evidences of gross environmental contamination and resultant ecological destruction wreaked by inundation of the biosphere with a veritable flood of synthetic alien molecules, chief among which must rank the conventional synthetic organic insecticides. The role of these during emergence of Green Revolution and thereby in meeting the ever-increasing demand for food and fibres to support ever-increasing world population is incontestable. However, a plethora of drawbacks, disadvantages, undesirable side effects and hazards stemming from indiscriminate and extensive application of conventional insecticides have surfaced during the last few decades (Sharma, 1982). These include phenomena such as pest flareback, secondary pest resurgence, resistance, broad spectrum, non-specific toxicity extending to non-target organism, persistence leading to universal distribution, biomagnifications etc. (Jacobson, 1990; Dhaliwal and Arora, 2003). To overcome the problems, the natural resources especially the botanical ones, offer unlimited scope towards the realization of this desideratum (Murugan and Babu, 1998).

Botanicals provide significant clues to Nature's own systems of checks and balances, which can be simulated to manage insect population. Thus, there exist a special category of chemicals variously termed secondary metabolites, substances, allelochemics, token stimuli, etc. (Sharma and Nagasampagi, 1979). The main function of these chemicals may profoundly influence the behaviour, physiology, growth, reproduction, development and nutrition of an insect pest.

In the present survey, we have compiled 128 exotic and 358 indigenous plants (Table 3.1) from 101 families (Table 3.2), because of

i.   reported references and cross references in ancient literature (Chopra, *et al.*, 1956, 1986; Frank Beye, 1978; Jacobson, 1990, Saxena and Shrivasta, 1993; Dhaliwal and Arora, 2003, etc.)

ii.   their folkloric reputation as insecticidal, repellant, antifeedant, attractant, growth disrupter and ovicidal properties (Jacobson and Crosby, 1971; Satyavati, 1976; Arora and Dhaliwal, 1994, etc.)

iii. their known ingredients like rotenone, saponin, alkaloids, nicotine, essential oils, pyrethrins, etc. (Chopra, *et al.*, 1986; Saxena and Shrivasta, 1993; Caius, 1998; Dhaliwal and Arora, 2003, etc.)

In the following list of plants, out of 101 families, some have more than 10 insecticidal plants, e.g. the higher numbers of plants were reported from Leguminosae (48), Asteraceae (36), Labiatae (34), Rutaceae (23), Solanaceae (22), Euphorbiaceae (20), Verbenaceae (14) and Myrtaceae (14). This indicates that the order of importance stated by earlier workers (Chopra *et al.*, 1956; Frank Beye, 1978; Jacobson and Crosby, 1971; Dhaliwal and Arora, 2003) is confirmed.

TABLE 3.1   LIST OF SUGGESTED POSSIBLE INSECTICIDAL PLANTS

No.	Botanical and common name	Family	Part used	Biologically active ingredient(s)	Mode of activity
Exotic Plants					
01	*Aconitum chinense* Linn Monkshood	Ranunculaceae	Wp	Alkaloid, aconitine	Insecticidal, antifeedant
02	*Aconitum napellus* Linn Mithazahar	Ranunculaceae	Rt	Alkaloid, aconitine, neopelline	Insecticidal
03	*Aesculus californica* Linn California buckeye	Hippocastanaceae	N, L Sd	Coumarins, saponin	Insecticidal
04	*Aframomum elegueta* K.Schum Melegueta pepper	Zingiberaceae	L	Paradol (Gingerol)	Insecticidal
05	*Alpinia galangal* Willd. Greater galangal	Zingiberaceae	Rh, O	Kaempferia, galangin, galangol	Insecticidal
06	*Amianthium muscaetoxicum* (Walt.). Fly poison	Liliaceae	Bb, L	Alkaloid	Insecticidal
07	*Amorpha fruticosa* Indigo bush	Araceae	Fr	Rotenoid	Insecticidal
08	*Anabasis aphylla* Linn	Chenopodiaceae	St, L	Anabasine	Insecticidal
09	*Anethum graveolens* Linn Dill	Apiaceae	L, Sd	Volatile oil, carvone, limnene	Repellent
10	*Annona senegalensis* Linn	Annonaceae	L	Alkaloid, resin, essential oil	Insecticidal
11	*Artemisia absinthium* Linn. Warm wood	Asteraceae	Wp	Anabsinthin, volatile oil, bitter glucoside	Insecticidal

12	*Boscia senegalensis* Pers.	Capparidaceae	L, Fr	Methyl isothiocyanate	Insecticidal
13	*Boswellia sacra* Flueck Luban	Burseraceae	Gum	Essential oil	Repellent, insecticidal
14	*Canna generalis* Canna lily	Musaceae	Be	Picrotoxin	Insecticidal
15	*Celastrus angulatus* Bitter sweet	Celastraceae	Rt, L, Br	Alkaloid	Insecticidal
16	*Cestrum parqui* L. Herit Green cestrum	Solanaceae	L	Saponin, essential oil, bitter aromatic substance	Antifeedant
17	*Chenopodium album* Linn Vastu	Chenopodiaceae	Wp	Essential oil, carvone and vit C.	Insecticidal
18	*Chenopodium ambrosioide* Linn. Jerusalem oak	Chenopodiaceae	Wp	Essential oil, saponin ascariodole, glycoside	Insecticidal
19	*Chrysanthemum balsamita* Linn Pyrethrum	Asteraceae	Fl	Pyrethrin-I and Pyrethrin-II	Insecticidal, repellent, antifeedant
20	*Chrysanthemum cinerariaefolium* Vis Pyrethrum	Asteraceae	Fl	Pyrethrin-I and Pyrethrin-II	Insecticidal, repellent, antifeedant
21	*Chrysanthemum coccineum* Willd (=*Pyrethrum roseum*)	Asteraceae	Fl	Pyrethrin-I and Pyrethrin-II	Insecticidal
22	*Chrysanthemum coronarium* Linn. Guldaudi	Asteraceae	Br, Fl	Adenine, chlonine	Anti-JH
23	*Chrysanthemum indicum* L. Manzanilla	Asteraceae	Fl, O	Essential oil	Insecticidal
24	*Cinnamomum verum* Presl. (Cinnamon)	Lauraceae	Br	Cinnamom oil—eugenol, essential oil	Growth disruptor

(Contd.)

TABLE 3.1   (CONTINUED)

No.	Botanical and common name	Family	Part used	Biologically active ingredient(s)	Mode of activity
Exotic Plants					
25	*Cissampelos owariensis* Linn Velvet leaf	Menispermeaceae	Rt	Hayatin, hayatinin, L-curine	Antifeedant
26	*Conium maculatum* Linn Jeera kala	Apiaceae	Sd	Alkaloid, D-coniline	Insecticidal
27	*Crotolaria juncea* Linn Kheshkhash	Leguminosae	Sd	Poisonous alkaloids	Repellent, insecticidal
28	*Cynanchum auriculatum* Wight Bhankalink	Asclepiadaceae	L	Toxic principles	Growth disruptor, insecticidal
29	*Delphinium ajacis* Linn Field larkspur	Ranunculaceae	Sd	Alkaloid, aconite	Insecticidal
30	*Delphinium delavayi* Larkspur	Ranunculaceae	Wp	Alkaloid, bitter substances	Insecticidal, repellent
31	*Delphinium staphisagria* Linn Lousewort	Ranunculaceae	Sd	Alkaloid—Aconite	Insecticidal
32	*Derris elliptica* (Roxb) Derris / Tabah	Leguminosae	Rt	Rotenone, resin—toxicol, tephrosine	Insecticidal (Larvicidal)
33	*Derris ferruginea* Benth Aru	Leguminosae	Rt	Rotenone	Insecticidal
34	*Derris trifoliata* Lour (Syn. *D. uliginosa*) Kirtana	Leguminosae	Br	Tannic acid, resin	Insecticidal
35	*Dryopteris filix-mas* L. Male fern	Polypodiaceae	Rh	Felicin	Insecticidal

36	*Duboisia leichhardtii* F. Mueller Pituri	Solanaceae	St, L	Nornicotine	Insecticidal
37	*Ecballium elaterium* L. A. Rich Kateri-indrayan	Cucurbitaceae	Fr	Glucoside, elaterin, ecballin, prophetin	Insecticidal
38	*Embelia ribes* Burm F.	Myrsinaceae	Sd	Embelin, quercitol, alkaloid, tannin	Insecticidal
39	*Eragrostis major* Host.	Cyperaceae	St, L	Bitter substances	Insecticidal
40	*Erythrophleum suaveolens* Brenan Red water tree	Caesalpiniaceae	Br	Alkaloid—Cassidine	Ovicidal, insecticidal
41	*Eucalyptus citriodora* H. Lemon-scented gum	Myrtaceae	O	Volatile oil—citronellal	Fumigant, insecticidal
42	*Eucalyptus hybrida*	Myrtaceae	L	Essential oil, terpenoids	Repellent, insecticidal
43	*Eucalyptus pauciflora* Sieb	Myrtaceae	L	Essential oil, terpenoids	Repellent, antifeedant
44	*Eucalyptus saligna* Sm	Myrtaceae	L	Essential oil	Antifeedant, repellent
45	*Eucalyptus terreticomis*	Myrtaceae	L	Essential oil	Antifeedant, repellent
46	*Euphorbia helioscopia* Linn	Euphorbiaceae	Sd, Ju	Saponin	Growth disruptor
47	*Euphorbia lateriflora* Schum and Thonn	Euphorbiaceae	St	Euphorbin, resin	Insecticidal
48	*Fortunella hindsii* Champ. Hong-kong kumquat	Rutaceae	Wp	Xanthotoxin	Insecticidal
49	*Gaultheria procumbens* L. Boxberry/Teaberry	Ericaceae	O	Methyl salicylate, gaultheric acid	Insecticidal
50	*Haematoxylon campechianum* L. Logwood tree	Leguminosae	Hw	Tannin	Insecticidal
51	*Haplophyton cimicidum* D.C.	Apocynaceae	St, L	Alkaloids	Insecticidal

(*Contd.*)

No.	Botanical and common name	Family	Part used	Biologically active ingredient(s)	Mode of activity
Exotic Plants					
52	*Helleborus niger* Linn Khorasani	Ranunculaceae	Rh	Helleborin, helleborein and hellebrin	Insecticidal
53	*Heliopsis scabra* Linn Oxeye	Boraginaceae	Rt	Scrabin	Insecticidal
54	*Heliopsis indicum* Linn	Boraginaceae	Rt	Scrabin	Insecticidal
55	*Heliotropium peruvianum* L. Heliotrope	Boraginaceae	Rt	Heliotropine	Insecticidal
56	*Hyoscyamus niger* Linn Ajowan	Solanaceae	Sd, L	Alkaloid, atropine, hyoscyamine	Insecticidal
57	*Hyptis spicigera* Lam Black sesame	Labiatae	Wp	Essential oil	Repellent, insecticidal
58	*Hyptis suaveolens* Poit	Labiatae	Wp	Essential oil	Repellent, insecticidal
59	*Jatropha dhofarica* (Zebrot) Linn Coral plant	Euphorbiaceae	L	Saponin, resin, tannin and bitter substances	Insect repellent, insecticidal
60	*Juglans regia* Linn Akroda	Juglandaceae	Fr	Alkaloid, barium, oxalic acid	Insecticidal
61	*Laurus nobilis* Linn. Sweet bay/Laurel	Lauraceae	O	Volatile oil—piperidine, geraniol	Repellent
62	*Lavendula angustifolia* Mill Lavender	Labiatae	Oil	Volatile oil, tannin, coumarin	Repellent

63	*Lippia geminata* HBK Wild sage	Verbenaceae	L	Alkaloid	Insecticidal
64	*Lonchocarpus utilis* Cube/Barbasco	Leguminosae	Rt	Rotenone	Insecticidal
65	*Lonchocarpus kunthnicou* Stinkwood	Leguminosae	Rt	Rotenone	Insecticidal
66	*Lupinus albus* Linn. Turmuz	Leguminosae	Sd	Alkaloid, lupinine, lupindine and lupamine	Insecticidal
67	*Lychnis coronaria* Desr Dusty miller	Caryophyllaceae	Wp	Alkaloids, tannin, bitter substances	Insecticidal, repellent, ovicidal
68	*Mammea americana* L. Mammey	Guttiferaceae	Sd, L	Mammein	Insecticidal
69	*Melanthium virginicum* Bulb Bunch flower	Sapindaceae	L	Bitter substances	Insecticidal
70	*Mentha longifolia* Linn Horse mint	Labiatae	O	Essential oil	Insecticidal
71	*Mentha piperita* Linn Peppermint	Labiatae	O	Essential oil	Repellent, insecticidal
72	*Milletia pachycarpa* B. Fish poison climber	Leguminosae	Rt, Sd	Alkaloids, rotenone, glysoside	Insecticidal
73	*Monodora myristica* Dunal Jamaican Nutmeg	Annonaceae	Sd	Essential oil—terpenes, alkaloid	Ovicidal and larvicidal
74	*Myrtus communis* Linn Yas	Myrtaceae	Fr, L	Essential oil	Repellent, insecticidal
75	*Nicotiana glunosa* Linn Tobacco	Solanaceae	L	Glucoside, alkaloid, nicotine	Insecticidal

(Contd.)

TABLE 3.1 (CONTINUED)

No.	Botanical and common name	Family	Part used	Biologically active ingredient(s)	Mode of activity
Exotic Plants					
76	*Nicotiana rustica* Linn Tobacco	Solanaceae	L	Glucoside, alkaloid, nicotimine	Insecticidal
77	*Nicotiana yustinue* Linn Tobacco	Solanaceae	L	Glucoside, anabasine, nicotine	Insecticidal
78	*Ocimum americanum* Linn Basil	Labiatae	St, L	Essential oil, carvacrol	Insecticidal
79	*Ocimum canum* Sims American basil	Labiatae	L	Volatile oil—linalool oil	Insecticidal
80	*Ocimum kilimandscharicum* Baker African basil	Labiatae	St, L	Essential oil, camphor	Insecticidal
81	*Ocimum suave* Willd Wild basil	Labiatae	O	Eugenol and mono- and sesquiterpenoids	Repellent
82	*Ocimum viride* Willd. Basil	Labiatae	St, L	Essential oil	Insecticidal, repellent
83	*Olea europaea* Linn Olive tree	Oleaceae	O	Cinchonidine, cinchonine, etc.	Insecticidal
84	*Origanum vulgare* Mill Sathra	Labiatae	Wp	Glucoside, volatile oil, quercetin	Insecticidal
85	*Pachyrhizus erosus* (L.) Urb. Yam Bean	Leguminosae	Sd	Pachyrhizin, dolineone, erismin	Insecticidal
86	*Pachyrhizus tuberosus* Yam Bean	Leguminosae	Sd	Pachyrhizin, dolineone, erismin	Insecticidal
87	*Peganum harmala* Linn Harmala	Rutaceae	Wp	Alkaloid—harmine, harmaline, peganine	Growth disruptor

88	*Pennisetum glaucum* (L.) R. Br Bajra	Poaceae	Wp	Amylase	Insecticidal
89	*Physalis mollis* Nutt. Field ground cherry	Solanaceae	St, L	Glycoside	Insecticidal
90	*Pogostemon heyneanus* Bth.	Labiatae	L, Rt, Fl	Essential oil	Repellent
91	*Populus candicans* Ait Balm of gilead	Salicaceae	Wp	Glucoside, salicin, bitter substances	Antifeedant, insecticidal
92	*Quassia amara* Linn Surinam quassia wood	Simaroubaceae	Rt, L	Amaroid, quassin and neoquassin	Larvicidal, insecticidal
93	*Quassia africana* Baill Quassia wood	Simaroubaceae	Rt, L	Quassin and neoquassin	Antifeedant
94	*Rhododendron hunnewelliana* Hook F Nao-yang-wha	Ericaceae	Fl	Andromedatoxin	Insecticidal
95	*Rhododendron molle* G. Don Yellow azalea	Ericaceae	Fl	Andromedatoxin	Insecticidal
96	*Rosmarinus officinalis* Linn Rosemary	Labiatae	Wp	Essential oil	Growth disruptor
97	*Ruta chalepensis* Linn Garden rue	Rutaceae	Wp	Glucoside, rutin, essential oil	Insecticidal, repellent
98	*Ryania speciosa* Vahl Ryania	Flacourtiaceae	Wp	Ryanodine	Feeding-deterrent, repellent, insecticidal
99	*Salvadora oleoides* Dcne	Salvadoraceae	L	Alkaloid, essential oil	Insecticidal, repellent
100	*Sambucus nigra* Linn Uti-Khaman	Caprifoliaceae	Br, Rt Be	Glucoside, sambunigrin, alkaloid, eldrin	Insecticidal
101	*Sapindus marginatus* Willd. Soapberry	Sapindaceae	Fr	Saponin	Insecticidal
102	*Salvia officinalis* Linn Sage/Sauge	Labiatae	O	α-pinene, limonene, camphor, terpineol	Fumigant, insecticidal

(Contd.)

TABLE 3.1 (CONTINUED)

No.	Botanical and common name	Family	Part used	Biologically active ingredient(s)	Mode of activity
Exotic Plants					
103	*Sophora alopecuroides* L.	Leguminosae	Sd, Rt	Alkaloid—cytosine	Insecticidal
104	*Sophora flavescens* Ait Yellow sophora	Leguminosae	Rt	Sparteine alkaloid	Insecticidal
105	*Sophora pachycarpa* Linn Sophora	Leguminoceae	Rt, L	Sparteine alkaloid	Insecticidal
106	*Schleichera trijuga* Willd. Lac tree	Sapindaceae	Sd	Phenol, polyene pigment, oxalic acid	Antifeedant
107	*Schoennocaulon officinale* Grey Sabadilla	Liliaceae	Sd	Alkaloid	Insecticidal, repellent
108	*Shorea robusta* Gaertn F. Sal-tree	Dipterocarpaceae	Wp	Resin (detergent), essential oil	Antifeedant
109	*Stellera chamaejasme* L. Lang-tu	Caryophyllaceae	Rt	Alkaloids	Insecticidal
110	*Stemona tuberosa* Lour Paipu	Roxburghiaceae	Rt	Alkaloid	Insecticidal
111	*Strychnos nux-vomica* Linn Strychnine tree	Loganiaceae	Rb	Strychnine, brucine	Insecticidal
112	*Suaeda aegyptiaca* Hasselq. Suwwad	Chenopodiaceae	L	Essential oil, glucoside	Repellent, insecticidal
113	*Tagetes patula* Linn Marigold	Asteraceae	Fl, L	Essential oil, bitter subs. sesquiterepens	Insecticidal
114	*Tephrosia virginiana* L. Pers Devil's shoestring	Leguminosae	Sd, Rt	Rotenoid, glucoside rutin	Insecticidal, antifeedant

115	*Tephrosia vogelii* Hook F. Vogel Tephrosia	Leguminosae	Sd, Rt	Rotenone	Insecticidal
116	*Tetradenia riparia* Hochst Umuruvumba	Labiatae	L	Linalool, diterpenol	Oviposition deterrent
117	*Thymus vulgaris* Linn	Labiatae	Fl, L	Essential oil, pinene, carvacrol, cymene	Growth disruptor
118	*Tithonia diversifolia* Hemsl. A. Gray Mexican sunflower	Asteraceae	Fl	Essential oil	Insecticidal
119	*Thymus vulgaris* Linn Garden thyme	Labiatae	O	Essential oil—thymol	Insecticidal
120	*Tripterygium wilfordii* Hook F. Thunder god vine	Celastraceae	R Br	Alkaloid, wilfordine	Insecticidal
121	*Urginea maritima* Kunth. Jangli piyaz	Liliaceae	Bb	Glycoside, scillaren A and B	Insecticidal
122	*Veratrum album* Linn. White hellebore	Liliaceae	Rh, Rt	Alkaloid—proveratrine A and B	Insecticidal
123	*Veratrum nigrum* Linn. Black hellebore	Liliaceae	Rt	Alkaloids	Insecticidal
124	*Vicia faba* Linn. Bakla	Leguminosae	Fr	Convicine, vicine	Insecticidal
125	*Xanthoxylum clava-herculis* Linn. Prickly ash	Rutaceae	Br	Alkaloid	Insecticidal
126	*Zanthoxylum bungeanum* Linn.	Rutaceae	Br, Fr	Essential oil, alkaloid	Insecticial, repellent
127	*Zanthoxylum hamiltonianum* Wall Hamilton prickly ash	Rutaceae	Rt	Essential oil, alkaloid, D-terepin	Larvididal, insecticidal
128	*Zanthoxylum zanthoxyloides* Lam.	Rutaceae	Fr	Alkaloid, essential oil	Repellent, insecticidal

(Contd.)

TABLE 3.1 (CONTINUED)

No.	Botanical and common name	Family	Part used	Biologically active ingredient(s)	Mode of activity
Indigenous plants					
01	*Abies balsamea* Linn. Miller	Pinaceae	L	Juvabione, dehydrojuvabione	Juvenile hormone analogue
02	*Abrus precatorius* Linn Gunj	Leguminosae	Sd	Abrin, glucoside, toxalbumin	Insecticidal
03	*Acacia arabica* Lamk. Babul	Leguminosae	G	Tannin, resin	Antifeedant
04	*Acacia concinna* DC Shikakai	Leguminosae	Pd	Saponin, alkaloid	Insecticidal
05	*Acacia nilotica* Linn Wild babul	Momosaceae	Br, L	Tannin, polyphenols, quercetin	Insecticidal
06	*Acalypha indica* Linn Khokali	Euphorbiaceae	Rt, L	Acalyphine, glucoside, HCN	Insecticidal
07	*Acorus calamus* Linn Sweet flag /Vekhand	Araceae	Rh	Glucoside, acorine, alkaloid, essential oil	Insecticidal, repellent, antifeedant
08	*Acorus gramineus* Soland Bhadra	Araceae	Rst	Essential oil	Insecticidal
09	*Adhatoda vasica* Nees Adulsa	Acanthaceae	L	Alkaloid—vasicine, essential oil	Antifeedant, insecticidal
10	*Adhatoda zeylanica* Medik Vasaka	Acanthaceae	L	Alkaloid—vasicolin, vasicine	Antifeedant, repellent
11	*Adina cordifolia* Roxb Haldu	Rubiaceae	B	Tannin, bitter substances	Juice used to kill maggots

12	*Aegiceras corniculatum* Blanco Halsi	Myrsinaceae	Br	Saponin, resin	Larvicidal
13	*Aegle marmelos* Corr. Bel	Rutaceae	R. Br, Fr	Marmalosin, alkaloid, umbelliferone	Repellent
14	*Aerva lanata* Juss Kapuri madhuri	Amaranthaceae	Rt, Fl, L	Flavonoid, glucoside, α-amyrine, β-sitosterol	Repellent
15	*Agave Americana* Linn American aloe	Amaryllidaceae	Wp	Volatile oil, saponin	Ravages of white ants
16	*Aglaia ganggo* Miq. Ganggo	Meliaceae	Wp	Alkaloid, glucoside	Insect resistant
17	*Aglaia odorata* Lour Mock lime	Meliaceae	F, L	Rocaglamide	Insecticidal
18	*Aglaia odoratissima* Blume	Meliaceae	Wp	Roxburghilin	Feeding deterrent
19	*Ageratum conyzoides* L. Goat weed	Asteraceae	Rt, L	Essential oil, alkaloid and coumarin	Anti-JH, ovicidal, repellent
20	*Ageratum houstonianum* Mill. Blue billy goat weed	Asteraceae	Wp	Precocene I, II, anacylin	Anti-JH activity
21	*Alianthus excelsa* Roxb Maharukh	Simaroubaceae	Br, L	Bitter substances ailanthin	Antifeedant, repellent
22	*Ajuga decumbens* Thunb Creeping bugleweed	Labiatae	L	Bitter substances, ajugalactons	Antifeedant, antiecdysone
23	*Ajuga remota* Benth Nilkanthi	Labiatae	L	Ajugarin I, II, III; β-ecdysone	Feeding deterrent
24	*Ajuga reptans* Linn Carpet bugle	Labiatae	L	Ajugareptansone A and B	Deterrent, development disruptor

(Contd.)

TABLE 3.1 (CONTINUED)

No.	Botanical and common name	Family	Part used	Biologically active ingredient(s)	Mode of activity
	Indigenous plants				
25	*Ajuga iva* (L.) Schreber French ground pine	Labiatae	L	Ajugapitin, clero-dane, diterpenoid	Feeding deterrent
26	*Albezia lebbeck* Wild sirissa	Leguminosae	Sd, L Po,Br	Alkaloid, caffeic acid and quercetin	Insecticidal
27	*Albizzia procera* (Roxb) Kinhai /Safed siris	Leguminosae	L	Saponin	Insecticidal
28	*Aleurites fordii* Hem Tung tree/Jangli akhrot	Euphorbiaceae	Sd	α-Eleostearic acid, octadecanol acetate	Oviposition-deterrent
29	*Allium cepa* Linn Onion	Liliaceae	Bb	Essential oil, quercetin—flavonoid	Oviposition-deterrent, antifeedant
30	*Allium sativum* Linn Garlic	Liliaceae	Bb	Diallyl sulphide, diallyl trisulphide	Repellent, antifeedant
31	*Aloe vera* Tourn Indian aloe	Liliaceae	Wp	Aloin, isobarbaloin, gum, resins	Insecticidal, antifeedant
32	*Alpinia galangal* Linn Kulanjan	Zingiberaceae	Rh	Essential oil	Insecticidal
33	*Anacardium occidentale* Linn. Cashew	Anacardiaceae	Br	Phenolic compound cardol, anacardic acid	Repellent, insecticidal
34	*Anacyclus pyrethrum* DC Pellitory	Asteraceae	Wp	Pellitorine/pyrethrin, anacyctin	Ovicidal, insecticidal
35	*Anagallis arvensis* Linn Jonk mari	Primulaceae	Wp	Saponin, glucoside	Insect repellent

36	*Anamirta cocculus* Linn Kakmari	Menispermeaceae	Wp	Picrotoxin, cocculin, anamirtin	Insecticidal
37	*Andira araroba* Aguiar Goa powder	Leguminosae	St	Chrysophanic acid	Insecticidal
38	*Andrachne cardifolia* Muell. Gurguli	Euphorbiaceae	L	Hydrocyanic acid	Insecticidal
39	*Andrographis paniculata* Nees Kalmegh	Acanthaceae	Wp	Kalmeghin, bitter substances	Insecticidal
40	*Andropogan schoenanthus* L. Lemon grass	Poaceae	Wp	Essential oil—geraniol	Repellent, antifeedant
41	*Annona muricata* Linn Soursop	Annonaceae	L	Alkaloid, essential oil	Insecticidal
42	*Annona reticulata* Linn Ramphal	Annonaceae	Br, L	Anonarine	Insecticidal
43	*Annona squamosa* Linn Sitaphal	Annonaceae	Sd, L	Amorphous alkaloid, toxic resin and oil	Insecticidal, repellent, antifeedant
44	*Arachis hypogaea* Linn. Ground nut	Leguminosae	Sd	Arachin, conarachin, nicotinic acid	Antifeedant, oviposition deterrent
45	*Areca catechu* Linn Supari/Betel nut	Palmae	Fl	Alks-arecaine, arecaidine, arecotine	Insecticidal
46	*Argemone mexicana* Linn Prickly poppy	Papaveraceae	Rt, Sd	Alkaloid, berberine, protopine	Larvicidal
47	*Arisaema speciosum* Wall Kiralu/Cobra lily	Araceae	Rt	Acrid juice	Insecticidal
48	*Arisaema tortuosum* Wall Whipcord cobra lily	Araceae	Sd, Rt	Acrid juice	Insecticidal

(*Contd.*)

No.	Botanical and common name	Family	Part used	Biologically active ingredient(s)	Mode of activity
Indigenous plants					
49	*Aristolochia bracteata* Wight Kidamari	Aristolochiaceae	L	Volatile substances, alkaloids	Larvicidal
50	*Artemisia capillaris* Thunb. Wormwood	Asteraceae	Rt, L	Capillin, capillarin	Feeding deterrent
51	*Artemisia vulgaris* Linn Indian Wormwood/Nagadona	Asteraceae	Rt	Essential oil, borneal, α-thujone	Repellent, insecticidal
52	*Artocarpus heterophyllus* Lam. Phanas	Moraceae	Rt, L	Tannin, steroketone, fixed oil	Oviposition-deterrent insecticidal
53	*Avicennia officinalis* Linn Bina	Verbenaceae	Rt, Sd	Tannin	Larvicidal
54	*Azadirachta indica* A. Juss Neem	Meliaceae	Br, L, Fr	Limnoid, azadirachtin, margosine	Antifeedant, oviposition-deterrent, insecticidal
55	*Balanites aegyptiaca* Linn Ingudi/Hingan	Simaroubaceae	Br, L, Fl	Saponin, glycoside, sapogenin	Repellent
56	*Bambusa arundinacea* (Retz.) Willd Spiny bamboo	Poaceae	Sh, L	Benzoic acid, cyanogenetic glucoside	Larvicidal (Mosquito)
57	*Barringtonia asiatica* Linn. Sea poison tree	Lecythidiaceae	Fr, L	Saponin, glucoside	Insecticidal
58	*Barringtonia racemosa* (L) Spreng. Powderpuff tree	Lecythidiaceae	Fr	Glucoside, saponin, tannin	Insecticidal
59	*Bignonia capreolata* Linn Trumpet flower	Bignoniaceae	L	Bitter substances, alkaloids	Ovicidal

60	*Bixa orellana* Linn Sinduri	Buxaceae	Wp	Triterepenes, essential oil bixin	Insect repellent
61	*Blumea balsamifera* (Linn) DC Camphor tree	Asteraceae	L, O.	Essential oil—borneol, cineole	Repellent, insecticidal
62	*Blumea lacera* DC. Blumee	Asteraceae	L, Fl	Essential oil	Repellent, antifeedant
63	*Bolusanthus speciosus* (Bolus) Harms Wisteria tree	Leguminosae	St	Bitter substances	Insecticidal
64	*Brassica campestris* Linn Field mustard	Cruciferae	Sd, Rt	Glyceride—crucic acid	Antifeedant, repellent
65	*Brassica juncea* Linn Brown mustard	Cruciferae	Sd, L	Essential oil	Antifeedant, repellent
66	*Brassica napus* Linn Rutabaga	Cruciferae	Rt	Essential oil	Antifeedant, repellent
67	*Brassica nigra* (Linn) K. Black mustard	Cruciferae	Sd	Glucoside, essential oil	Antifeedant, repellent, insecticidal
68	*Brucea sumatrana* Roxb Java brucea	Simaroubaceae	Rt	Bitter substances	Insect repellent
69	*Bursera penicillata* (DC) Eng Linaloe tree	Burseraceae	Hw	Essential oil, linaloe oil	Antifeedant, insecticidal
70	*Butea frondosa* Roxb. Flame of the forest	Leguminosae	L, O	Essential oil	Antifeedant, insecticidal
71	*Butea monosperma* Lam Palas	Leguminosae	Fl, L	Glucoside, butrin, butein, butin, etc.	Insecticidal
72	*Caesalpinia pulcherrima* (L) Peacock flower	Leguminosae	Fl	Phytosterin, boducin, saponin	Insecticidal

(Contd.)

TABLE 3.1 (CONTINUED)

No.	Botanical and common name	Family	Part used	Biologically active ingredient(s)	Mode of activity
Indigenous plants					
73	*Calonyction muricatum* (Linn) G. Don Purple moonflower	Convolvulaceae	L	Resin, glucoside	Insecticidal
74	*Calophyllum inophyllum* Linn Alexandrian laurel	*Clusiaceae*	Wp	Glucoside, flavonol-myricetin, quercetin	Insecticidal
75	*Calotropis gigantean* N. Calotrope/crownflower	*Asclepiadaceae*	L	Catachin, α-calotropeol, β-calotropeol	Growth inhibitor
76	*Calotropis procera* (Ait.) Aak	*Asclepiadaceae*	L	Calotropin and calotropagenin	Ovicidal, larvicidal, antifeedant
77	*Canavalia ensiformis* Linn Jack bean	Leguminosae	Fr	Canavanine, canavaline	Insecticidal, metabolic disrupter
78	*Canna indica* Linn Wild canna lily	Musaceae	Fl	Canavaline	Insecticidal
79	*Cannabis sativa* Linn Bhang/Ganja	Cannabinaceae	L	Cannabinol, cannabinin, pseudocannabinol	Insecticidal
80	*Canarium commune* Linn Java almond tree	Burseraceae	Fr	Resin, essential oil, anethole and terpenes	Insecticidal
81	*Capparis decidua* (Forsk) Edgew Karer	Capparidaceae	Tw	Alkaloid	Insecticidal
82	*Capsicum annum* Linn Red Chilli/*Mirachi*	Solanaceae	Fr, L	Capsaicin	Antifeedant, insecticidal
83	*Capsicum frutescens* Linn Cayenne pepper	Solanaceae	Fr, L	Capsaicin and dihydrocapsaicin	Insecticidal

84	*Carapa granatum* (Koenig) Alston Dhundal	Meliaceae	Fr, Br Sd	Bitter tonic, tannin	Insecticidal, repellent
85	*Carapa procera* DC Monkey cola	Meliaceae	Wp	Carapolides, mexicanolide	Feeding deterrent
86	*Carica papaya* Linn Papaya/Papita	Caricaceae	L	Carpaine, papain	Repellent
87	*Carthamus tinctorius* Linn American saffron	Asteraceae	O	Carthamin	Insecticidal
88	*Carum carvi* Linn. Caraway/Jira	Apiaceae	Sd	Essential oil—carvon terpene and carvacrol	Antifeedant, repellent
89	*Carum roxburghianum* Benth Bishop's weed	Apiaceae	Sd	Essential oil—carvon terpene and carvacrol	Antifeedant, repellent
90	*Cassia didymobotrya* Frensen Candle bush, Senna	Leguminosae	Sd, L	Tannin, essential oil	Insecticidal, repellent
91	*Cassia fistula* Linn Indian Laburnum/Amaltas	Leguminosae	Wp	Anthraquinone, tannin	Insecticidal, ovicidal
92	*Cassia occidentalis* Linn Coffee senna	Leguminosae	L, Rt, Sd	Tannic acid, toxalbumin, fatty oil	Phagodeterrent
93	*Cassia siamea* Lam Kassod	Leguminosae	L, pd	Toxic alkaloids	Oviposition-deterrent
94	*Cassia tora* Linn Foetid cassia	Leguminosae	L, Rt	Emodin, glycoside, questin, β-sitosterol	Insecticidal
95	*Castanea dentata* Marsh Sweet chestnut	Fagaceae	Wp	Tannin	Insecticidal
96	*Casuarina equisetifolia* Linn Iron wood	Casuarinaceae	St	Tannin	Insecticidal

(*Contd.*)

TABLE 3.1 (CONTINUED)

Indigenous plants

No.	Botanical and common name	Family	Part used	Biologically active ingredient(s)	Mode of activity
97	*Cedrus deodara* Roxb (Deodar)	Pinaceae	Br, Hw	Gum, cholesterine, essential oil	Larvicidal
98	*Celastrus angulatus* Maxim Bitter sweet	Celastraceae	Rt	Quinine, essential oil, glycoside	Insecticidal, repellent
99	*Centratherum anthelmintcum* Willd (Somraj/kalijiri)	Asteraceae	Sd	Bitter substances	Insecticidal
100	*Cimicifuga foetida* Linn Bugbane	Ranunculaceae	Rh	Saponin, glucoside, tannin, essential oil	Repellent
101	*Cinnamomum aromaticum* N. Cassia	Lauraceae	Br	Cinnamomum, laurolisine	Repellent, insecticidal
102	*Cinnamomum camphora* Nees Camphor	Lauraceae	Br	Essential oil, camphor, campherol	Repellent, insecticidal
103	*Cinnamomum cecicodaphne* Nees Gonsoroi	Lauraceae	Br	Essential oil	Insecticidal, repellent
104	*Cinnamomum zeylanicum* Blume Cinnamon	Lauraceae	Br	Essential oil, eugenol	Insecticidal, repellent
105	*Cissampelos owariensis* P. Beauv Pareira Brava	Menisper-meaceae	L, Rt	Alkaloid—chondrode-ndrine, berberine, etc.	Insecticidal
106	*Cissus quadrangularis* Linn Veld grape	Vitaceae	Wp	Calcium oxalate, carotene	Repellent
107	*Citrullus colocynthis* Shrad Colocynth	Cucurbitaceae	Fr	Colocynthin, tannin, glucoside, saponin	Insecticidal, antifeedant

No.	Plant species	Family	Part	Chemical constituents	Activity
108	*Citrus aurantifolia* Swingle Lime/Lima	Rutaceae	Fr	Volatile oil—limonene	Feeding deterrent, insecticidal
109	*Citrus aurantium* Linn Sour orange	Rutaceae	Fr	Linalool	Feeding deterrent, oviposition-deterrent
110	*Citrus grandis* Linn Pomelo	Rutaceae	L, Fr	Essential oil, oil of lemon	Oviposition-deterrent
111	*Citrus limon* Linn Lemon	Rutaceae	Fr	Limonin, nomilin, obacunone	Insecticidal, antifeedant
112	*Citrus paradise* Macf Grape fruit	Rutaceae	Fr	Isolimnoic acid, limonin	Insecticidal, oviposition-deterrent
113	*Citrus sinensis* (Linn) Sweet orang/Musambi	Rutaceae	Fr	Nomilin, harrisonin, D-limonene, essential oil	Antifeedant, insecticidal
114	*Citrus medica* Linn Citron	Rutaceae	Fr, Ju, Rt	Peel of limonene, dipentene, citral	Larvicidal, repellent
115	*Cleome icosandra* Linn Wild mustard	Capparidaceae	St, L	Viscoid acid, viscosin	Antifeedant, repellent
116	*Cleome monophylla* Linn	Cleomaceae	St, L	Essential oil—terpenolene	Antifeedant, repellent
117	*Clerodendron inerme* Linn Embrert, Indian privet	Verbenaceae	Rt, L	Bitter principles, resin, gum	Insecticidal
118	*Clerodendrum infortunatum* Linn Hill glory flower	Verbenaceae	Rt, L	Bitter substance clerodin, trandecalin	Antifeedant, insecticidal
119	*Clerodendrum multiflorum* (Burn F.) O. Ktze. Agnimantha	Verbenaceae	Wp	Alkaloid	Antifeedant, insecticidal
120	*Clerodendrum phlomidis* Linn Arni	Verbenaceae	L	Bitter substances	Antifeedant, insecticidal
121	*Clerodendrum serratum* Blue fountain bush	Verbenaceae	St	Alkaloid, glucoside-saponin, bitter subst.	Repellent, antifeedant

(Contd.)

TABLE 3.1 (CONTINUED)

No.	Botanical and common name	Family	Part used	Biologically active ingredient(s)	Mode of activity
Indigenous plants					
122	*Cocculus hirsutus* Linn Broom creeper	Menispermaceae	Wp	Toxic alkaloid—coclaurine	Insecticidal
123	*Cocos nucifera* Linn Coconut	Arecaceae	Fr, L	Oil contains enzymes, milk contains amino acids	Antifeedant
124	*Coleus amboinicus* Lour Indian borage	Labiatae	O	Potassium and ethereal oil and phenol	Insecticidal
125	*Colocasia esculenta* Linn Green taro	Araceae	Tuber	Sapotoxin	Insecticidal
126	*Corchorus capsularis* Linn Jute/white jute	Tiliaceae	Sd	Glucoside corchorin, capsularin	Insecticidal
127	*Coriandrum sativum* Linn. Coriander	Apiaceae	Sd	Essential oil—coriandrol	Repellent, antifeedant
128	*Croton gratissimum* Burch Lavender croton	Euphorbiaceae	L	Alkaloid—resin	Insecticidal, antifeedant
129	*Croton oblongifolius* Rox Chucka	Euphorbiaceae	Br, Rt	Fatty oil, resin	Insecticidal
130	*Croton tiglium* Linn Purging croton	Euphorbiaceae	Sd	Croton, resin, protein alkaloid, ricinine	Insecticidal
131	*Cucumis sativum* Linn Cucumber	Cucurbitaceae	Fr	Bitter substances	Larvicidal
132	*Cucurbita pepo* Linn Pumpkin/Kaula	Cucurbitaceae	Sd	Resin—phytostearin and salicylic acid	Antifeedant, insecticidal

133	*Curcuma longa* Roxb (Turmeric/Halad)	Zingiberaceae	Rh	Essential oil—linalool, ocimene, D-a-pinene	Insecticidal, antifeedant
134	*Cuminum cyminum* Linn Cumin	Apiaceae	Sd	Essential oil, cuminol	Insecticidal, antifeedant
135	*Curculigo orchioides* Gaertn Kali-Musali	Hypoxidaceae	Rh, L	Glycosides, alkaloid lycorine, yuccagenin	Antifeedant
136	*Cycus revoluata* Thumb Sago palm	Amaranthaceae	M C	Toxic glucoside	Insecticidal, attractant
137	*Cymbopogan citratus* D.C. West Indian Lemon grass/oil grass	Poaceae	Wp	Essential oil, citral, cymbopogonol	Repellent
138	*Cymbopogan flexuosus* (Steud.) Wats East Indian Lemon grass	Poaceae	L	Essential oil, citral	Repellent
139	*Cymbopogan martini* Roxb Rosha grass	Poaceae	L	Essential oil	Repellent
140	*Cymbopogan nardus* Linn Citronella grass	Poaceae	L	Essential oil, citronellae	Repellent
141	*Cyperus iria* Linn Umbel sedge	Cyperaceae	Wp	Essential oil—pinene, sesquiterpenes	Natural JH
142	*Cyperus rotundus* Linn Common nut sedge	Cyperaceae	Tb	Essential oil—pinene, cineole, sesquiterepene	Oviposition-deterrent, insecticidal
143	*Dalbergia retusa* Hemsl (Cocobolo)	Fabaceae	Wp	Obstaquinone, alkyl phenol	Repellent, growth inhibitor
144	*Dalbergia sissoo* Linn Indian rose wood	Fabaceae	Wp	Bitter principles, tannin	Ovicidal
145	*Datura metel* Linn Thorn apple	Solanaceae	L	Alkaloid, atropine, hyoscyamine	Repellent, insecticidal

(Contd.)

TABLE 3.1   (CONTINUED)

No.	Botanical and common name	Family	Part used	Biologically active ingredient(s)	Mode of activity
Indigenous plants					
146	*Datura strantonium* Linn Jimson weed	Solanaceae	Sd, L	Atropine, hyoscine, hyoscyamine	Reduce fecundity
147	*Delphinium branonianum* Royle Musk larkspur	Ranunculaceae	L	Alkaloid—delphinine, taphisagroine	Insecticidal (Ticks)
148	*Delphinium caeruleum* Jacq. Dhakanga	Ranunculaceae	Rt	Sagroine	Larvicidal (Maggots)
149	*Delphinium elatum* Linn Candle larkspur	Ranunculaceae	Sd	Alkaloid—delatine, delpheline	Insecticidal
150	*Delonix regia* Gamble Flame tree	Leguminosae	Fl	Alkaloid, bitter principles	Ovicidal
151	*Derris fordii* Olive Derris	Leguminosae	Rt	Rotenone, glucoside	Insecticidal
152	*Derris indica* Lam Pongame oil tree	Leguminosae	Rt	Rotenone	Insecticidal
153	*Dieffenbachia seguinae* Jack Sleahott/Dumbcane	Araceae	Wp	Calcium oxalate	Insecticidal
154	*Dioscorea bulbifera* L. Air-potato, Air yam	Dioscoreaceae	Tb	Poisonous glucoside	Insecticidal
155	*Dioscorea deltoidea* Wall Kirta/Kildri	Dioscoreaceae	Tb	Saponin	Larvicidal
156	*Dioscorea hispida* Dennst Potato yam	Dioscoreaceae	Tb	Alkaloid—dioscorine	Larvicidal (Maggot)

No.	Botanical name / Common name	Family	Part	Chemical constituents	Action
157	*Dioscorea prazeri* prain Wild yam	Dioscoreaceae	Tb	Saponin	Lousicidal
158	*Dodonaea viscosa* (Linn) Jacq Hop-bush	Sapindaceae	L	Alkaloid, saponin, glucoside	Antifeedant, repellent, insecticidal
159	*Duranta plumieri* Jacq Golden dewdrop	Verbenaceae	L	Saponin, alkaloid, narcotine	Larvicidal, insecticidal
160	*Duranta repens* Linn Duranta	Verbenaceae	L	Saponin	Insecticidal
161	*Dysoxylum malabaricum* Bedd Vellakil	Meliaceae	St	Bitter substances	Insecticidal
162	*Echinops echianthus* Roxb Indian globe thistle	Asteraceae	Rbr	Echinopsine	Insecticidal
163	*Echinochloa frumentacea* Link Millet/jungle rice	Gramineae	Wp, L	Bitter substances	Antifeedant
164	*Echinacea angustifolia* DC Hedgehog flower	Asteraceae	Wp	Echinolone	JHA
165	*Eichhornia crassipes* Solms Water hyacinth	Pontedariaceae	Rh	Carotene, bitter substances	Antifeedant
166	*Eleusine coracana* Linn Ragi, finger millet	Gramineae	L	Vitamin B, bitter substances	Antifeedant
167	*Ervatamia divaricata* Linn East Indian rosebay / Tagar	Apocynaceae	L	Alkaloid, caronarine and tabernaemontanine	Insecticidal
168	*Eruca vesicaria* Linn Gardenrocket	Cruciferae	O	Essential oil	Oviposition-deterrent
169	*Erythrina indica* Lam East Indian Coral	Leguminosae	Br, L, Sd	Poisonous alkaloid, saponin	Ovicidal

(Contd.)

TABLE 3.1 (CONTINUED)

No.	Botanical and common name	Family	Part used	Biologically active ingredient(s)	Mode of activity
	Indigenous plants				
170	*Eucalyptus globul* Labill Eucalyptus	Myrtaceae	L	Essential oil—cineole, euphorbine, sesquiterpene	Repellent, insecticidal
171	*Eugenia caryophyllus* Thunb Clove	Myrtaceae	Fl	Essential oil—eugenol	Antifeedant, insecticidal
172	*Euonymus europaeus* Linn Pegwood, spindle tree	Celastraceae	Fl, Br	Alkaloids	Insecticidal
173	*Eupatorium triplinerve* Vahl Ayapana tea	Asteraceae	Wp	Bitter principles inulin eupatorin and resin	Insecticidal
174	*Euphorbia antiquoram* Linn Triangular/square spurge	Euphorbiaceae	St, Lx	Euphorbin	Insecticidal
175	*Euphorbia caducifolia* Haines Cactus	Euphorbiaceae	Lx	Alkaloid	Repellent
176	*Euphorbia hirta* Linn Asthma weed	Euphorbiaceae	Wp	Alkaloid, essential oil	Larvicidal
177	*Euphorbia thymifolia* Linn Thyme-leaf spurge	Euphorbiaceae	Rt, Lx	Essential oil, 3-hydro flavone 7-glucoside	Insecticidal
178	*Euphorbia tirucalli* Linn Petrolium plant/milk bush	Euphorbiaceae	St, Lx	Euphorbone, ketone	Larvicidal
179	*Evolvulus alsinoides* Linn Dwarf moring glory	Convolvulaceae	L	Alkaloid—evolvine, α-sitosterol	Insecticidal
180	*Excoecaria agallocha* Linn Blinding tree	Euphorbiaceae	Br	Acrid latex, tannin	Insecticidal

181	*Ferula asafetida* Regel (Hing) Asafoetida	Apiaceae	Gm	Essential oil, resin, ferulic acid, pinene	Insecticidal
182	*Ficus benghalensis* Linn (Vad) Krishna fig	Moraceae	Fr. L	Hypoglycaemic glucoside, quercetin	Insecticidal
183	*Gardenia campanulata* R. Dikamali	Rubiaceae	Fr	Bitter substance	Larvicidal
184	*Gardenia lucida* Roxb Brilliant gardenia	Rubiaceae	Gm	Essential oil, resin, bitter substances	Repellent, larvicidal
185	*Gaultheria fragrantissima* wall Winter green	Ericaceae	L	Essential oil	Repellent, insecticidal
186	*Ginkgo biloba* Linn Ginkgo tree)	Ginkgoaceae	Wp	Salicylic acid, bilobalide	Feeding deterrent
187	*Gloriosa superba* Linn Glory lily/Tigerclaw	Liliaceae	Rt	Alkaloid—serpentine, gloriosine	Larvicidal
188	*Glycine max* (Linn) Merrill Soyabean	Fabaceae	Wp	Pinitol	Antifeedant, insecticidal
189	*Gossypium indicum* Linn Cotton	Malvaceae	SO	Quercimeristrin and isoquercitin	Anifeedant, insecticidal
190	*Gynandropsis gynandra* (L) Briquet Spider flower	Capparidaceae	Sd	Essential oil, cleomin	Insecticidal
191	*Hardwickia binata* C.B. Clarke Angian	Fabaceae	Wd	Mapanol and epicatechin	Antifeedant
192	*Helenium amurum* Raf. Bitter weed	Asteraceae	Wp	Helenalin, linifolin A., tenulin	Repellent, growth inhibitor
193	*Helianthus annus* Linn Sunflower	Asteraceae	Sd, L	Chlorogenic acid, carotin, lutein	Antifeedant, insecticidal

(*Contd.*)

TABLE 3.1   (CONTINUED)

No.	Botanical and common name	Family	Part used	Biologically active ingredient(s)	Mode of activity
Indigenous plants					
194	*Hibiscus rosa sinensis* Linn. Shoe flower	Malvaceae	St	Essential oil	Antifeedant, insecticidal
195	*Hibiscus syriacus* Linn Althea/Rose of sharon	Malvaceae	WP	Pelargonic acid, alkyl fatty acid	Oviposition- and feeding deterrent
196	*Hippocratea indica* Willd	Hippocrateaceae	Sd, L	Bitter substance, alkaloid	Lousicidal
197	*Hiptage benghalensis* Kurz Hiptage	Malpighiaceae	Rt, L	Glucoside—hiptagin	Insecticidal
198	*Holarrhena antidysenterica* Wall Koneru/Tellicherry bark	Apocynaceae	Sd, Br	Alkaloid—conessine, karchine, karchicine	Insecticidal
199	*Hura crepitans* Linn Sandbox tree	Euphorbiaceae	Wp	Toxic substances, crepitin and alkaloid	Insecticidal
200	*Hydnocarpus wightiana* Blume Marotti oil tree	Flacourtiaceae	Sd	Fatty oil hydnocarpic, chaulmoogric acid	Antifeedant
201	*Inula helenium* Linn Elecampane/Wild sunflower	Asteraceae	Rt	Essential oil, bitter substances	Repellent, antifeedant
202	*Inula rolekana* DC Himalayan elecampane	Asteraceae	Rt	Alkaloid	Insecticidal
203	*Ipomoea carnea* Jacq. Morning glory tree	Convolvulaceae	L	Bitter substances, alkaloids	Insecticidal
204	*Ipomoea fistulosa* Mart. Shrubby morning glory	Convolvulaceae	L	Glucoside—ipomoein	Insecticidal, ovicidal

205	*Ipomoea palmata* Linn Five-fingered morning glory	Convolulaceae	L	Glucoside, resin	Ovicidal
206	*Jasminum officinale* Linn Jasmine/Jai	Oleaceae	Fl	Lupeol, jasminol, flavonoids	Insecticidal
207	*Jatropha curcus* Linn Jatropha	Euphorbiaceae	Sd	Toxic substances, resin, curcin	Insecticidal
208	*Juglans regia* Roxb English walnut	Juglandaceae	L	Alkaloid	Repellent
209	*Juniperus oxycedrus* Linn Prickly juniper	Cupressaceae	Wd	Essential oil	Insecticidal
210	*Juniperus virginiana* Linn Eastern red ceder	Cupressaceae	Wp	Essential oil	Insecticidal
211	*Justicia adhatoda* Linn Malabar nut tree	Acanthaceae	L	Vascine, vasicinol, vasicione	Antifeedant
212	*Justicia gendarussa* Burm. F. Kuntze	Acanthaceae	Wp	Alkaloid	Insecticidal
213	*Kalanchoe integra* (Medik.) Kuntz Neverdie	Crassulaceae	Wp	Alkaloid, bitter substances	Antifeedant, insecticidal
214	*Kaempferia galanga* Linn Kapurkachali	Scitaminaceae	Wp	Alkaloid, essential oil	Insecticidal
215	*Kalanchoe spathulata* DC Haiza/Tatara	Crassulaceae	L	Essential oil	Insecticidal
216	*Kalmia latifolia* Linn Mountain laurel	Ericaceae	Wp	Kalmitoxin, grayanotoxin–III	Antifeedant
217	*Lagenandra toxicaria* Dalz Rukh-alu	Cucurbitaceae	L	Bitter substances	Insecticidal

(*Contd.*)

TABLE 3.1 (CONTINUED)

No.	Botanical and common name	Family	Part used	Biologically active ingredient(s)	Mode of activity
Indigenous plants					
218	*Lantana camera* Linn Lantana	Verbenaceae	L	Essential oil—camerene	Repellent, antifeedant
219	*Lavandula officinalis* Linn Common lavander	Labiatae	Wp	Essential oil	Larvicidal
220	*Lawsonia inermis* Linn. Mehandi	Lythraceae	L	Glucoside, gennotanic acid	Antifeedant, oviposition-deterrent
221	*Leucas cephalotes* Spreng Tumba	Labiatae	Wp	Essential oil, alkaloid	Insecticidal
222	*Leucas martinicensis* R.Br Wild tea bush	Labiatae	Wp	Essential oil	Repellent
223	*Limonia acidissima* Linn Wood apple	Rutaceae	Br, L	Bergapten, psoralen	Oviposition-deterrent, insecticidal
224	*Linum usitatissimum* Linn Linseed	Linaceae	SO	Volatile oil	Repellent
225	*Lobelia nicotionifolia* Wild tobacco	Campanulaceae	L	Alkaloid, lobelinic acid	Insecticidal
226	*Lufa acutangula* Roxb Ribbed sponge gourd	Cucurbitaceae	Fl	Crude saponin, cucurbitacin-B (amarin)	Insecticidal
227	Lycopersicon esculentum Mill Tomato	Solanaceae	L	Narcotine, solanine, oxalic acid	Insecticidal
228	*Madhuca longifolia* Linn Indian butter tree	Sapotaceae	Fl. Sd	Mowrin	Insecticidal

No.	Plant species / common name	Family	Part	Active principle	Activity
229	*Madhuca indica* Linn. Mahua	Sapotaceae	Fl	Alkaloid, saponin	Insecticidal, repellent
230	*Madhuca latifolia* Roxb. Mahua	Sapotaceae	Fl	Alkaloid, saponin	Insecticidal
231	*Maesa chisia* D. Don Susi-porma	Myrsinaceae	Rt, L, Br	Saponin	Insecticidal
232	Mangifera indica Linn Mango	Anacardiaceae	Fl	Flavones, essential oil, mangiferin	Insecticidal
233	*Maranta arundinacea* Linn Arrow root	Marantaceae	Rh	Bitter substances	Antifeedant
234	*Medicago sativa* Linn Alfalfa	Fabaceae	Wp	Butyric acid, succinic acid, coumarins	Repellent, insecticidal
235	*Melaleuca leucadendron* Linn White tea tree	Myrtaceae	Tw. L	Essential oil	Repellent
236	*Melia azedarach* Linn Lilae, pride of India	Meliaceae	Fr	Tetranortriterpenoids, meliandiol, etc.	Antifeedant, oviposition-deterrent
237	*Meliltous indica* All. Sweet clover	Leguminosae	Fl	Coumarin	Antifeedant
238	*Melinis minutiflora* (Beauv) Honey grass	Poaceae	Wp	Essential oil	Insect repellent
239	*Mentha arvensis* Linn Mint	Labiatae	L	Essential oil, carene, D-Carvone, citronellol	Antifeedant, repellent
240	*Mentha pulegium* Linn Pudding grass	Labiatae	Wp	α-pinene, β-pinene, essential oil	Insecticidal
241	*Mentha spicata* Linn Spearmint/garden mint	Labiatae	Wp	Carvone, pulegene, menthol	Antifeedant, insecticidal

(Contd.)

No.	Botanical and common name	Family	Part used	Biologically active ingredient(s)	Mode of activity
Indigenous plants					
242	*Mentha viridis* Linn Pudina	Labiatae	Wp	Essential oil	Insecticidal
243	*Nicandra physaloides* Shoo fly plant	Solanaceae	L	Nicandrone, nicandrine	Antifeedant
244	*Michelia champaca* Linn Champak	Magnoliaceae	Fl	Alkaloid—liriodenine	Insecticidal, repellent
245	*Mimusops elengi* Linn Spanish cherry	Sapotaceae	Br, L Sd	Quercitol, quercetin, β-sitosterol, saponin	Insecticidal
246	*Momordica charantia* Linn Bitter gourd	Cucurbitaceae	L	Charantin, foetidin, momordicin	Insecticidal, larvicidal
247	*Mundulea sericea* willd Alligator plant	Leguminosae	Br	Rotenoid, glucoside	Insecticidal
248	*Murraya exotica* Linn Orange jasmine	Rutaceae	L	Essential oil, glucoside, murrayin	Repellent, antifeedant
249	*Murraya koenigii* Spreng Curry leaf	Rutaceae	L	Essential oil, glucoside, koenigin	Repellent
250	*Musa sapientum* Linn Kadhali vaazha	Musaceae	L, Fr	Serotonin, dopamine and norepinephrine	Insecticidal
251	*Myristica fragrans* Houtt. Nutmeg	Myristicaceae	Fr, L	Eugenol, iso-eugenol, terpineol, linalool	Insecticidal
252	*Nephrolepis exaltata* Linn Fern	Polypodiceae	L	Alkaloids	Insecticidal

253	*Nerium indicum* (Miller) Arali	Apocynceae	L	Glucoside—toxic principles, neriodorin	Insecticidal
254	*Nerium odorum* Linn Sweet-scented oleander	Apocyanceae	L	Oleandrin, bitter substances	Insecticidal
255	*Nerium oleander* Linn Oleander	Apocynaceae	L	Cardiotonic, oleandrin, neridin	Antifeedant, insecticidal
256	*Nicotiana tabacum* Linn Tobacco	Solanaceae	L	Glucoside, nicotine, quercetin, anabasine	Contact insecticide, fumigant
257	*Nigella sativa* Linn Black cumin	Rananculaceae	Sd	Essential oil, glucoside, melanthin	Insecticidal
258	*Nyctanthus arbortristis* Linn Coral jasmine/Queen of the night	Oleaceae	Fl	Alkaloid, resin, glucoside	Attractant, insecticidal
259	*Ochna pulcra* Linn Peeling plane	Ochnaceae	L	Bitter principles, alkaloids	Antifeedant, ovicidal
260	*Ocimum basilicum* Linn Sabza /Basil	Labiatae	Wp	Essential oil, juvacimene-I, II	JHA, insecticidal, larvicidal
261	*Ocimum grutissimum* Linn Ramtulsi/Wild basil	Labiatae	Wp	Essential oil, thymol, eugenol	Insect (Mosquito) repellent
262	*Ocimum sanctum* Linn Tulsi/Sacred basil	Labiatae	L	Essential oil, eugenol, carvacrol	Repellent, antifeedant
263	*Ougenia dalbergioides* B. Chariot tree	Leguminosae	Br, L	Essential oil	Larvicidal
264	*Parthenium hysterophorus* Linn Congress grass	Asteraceae	L	Parthenin	Feeding deterrent, growth inhibitor
265	*Peganum harmala* Linn Syrian rue	Rutaceae	Wp	Alkaloid—harmine, harmaline, peganine	Insecticidal

(Contd.)

TABLE 3.1 (CONTINUED)

No.	Botanical and common name	Family	Part used	Biologically active ingredient(s)	Mode of activity
	Indigenous plants				
266	*Phyllanthus niruri* Linn Carry me seed	Euphorbiaceae	Wp	Bitter substances-phyllanthrin	Antifeedant, insecticidal
267	*Phytolacca acinosa* Roxb Indian poke berry	Phytolaccaceae	Rt	Bitter substances, phytolacca toxin	Insecticidal
268	*Pimenta acris* Kostel	Myrtaceae	L	Essential oil	Repellent, insecticidal
269	*Pimenta racemosa* J.W. Moore Bay rum tree	Myrtaceae	Be	Essential oil	Insecticidal
270	*Pimpinella anisum* Linn Sweet cumin	Apiaceae	Sd	Essential oil—pinene, limonene	Insecticidal
271	*Pinus taeda* Linn Loblolly pine	Pinaceae	Sd	Essential oil	Insecticidal
272	*Pinus virginiana* Mill Virginia pine	Pinaceae	Sd	β-pinene, pinene	Insecticidal
273	*Piper betal* Linn Betel pepper	Piperaceae	L	β-sitosterol, γ-sitosterol, glucoside	Insecticidal
274	*Piper cubeba* Linn. F. Tailed pepper	Piperaceae	SO	Essential oil, piperine, sylvatine	Oviposition-deterrent, insecticidal
275	*Piper guineense Schum* Ash pepper	Piperaceae	Fl	Essential oil, Cubebin	Repellent, insecticidal
276	*Piper longum* Linn Long pepper	Piperaceae	Fr, Fl	Piperine, sesquiterepenes	Insecticidal

277	*Piper nigrum* Linn Black pepper	Piperaceae	Fr, Rt	Piperine, amides	Insecticidal, oviposition-deterrent
278	*Plumbago capensis* Thunb Chitrtaka	Plumbaginaceae	Rt	Plumbagin	Antifeedant, chitin synthesis inhibitor
279	*Podophyllum peltarum* Linn May apple	Berberidaceae	Rh, Rt	Podophyllotoxin, podophyllin	Repellent, cytotoxin
280	*Pogostemon heyneanus* Benth Patchouli	Labiatae	Wp	Essential oil	Insect repellent
281	*Polyalthia longifolia* Benth Ashoka	Annonaceae	L	Alkaloid, toxic resin	Antifeedant
282	*Polygonum flaccidum* Meis. Dhakata sheval	Polygonaceae	Wp	Bitter substance	Larvicidal (Mosquito)
283	*Polygonum hydropiper* Lour Water pepper	Polygonaceae	St, L	Essential oil, oxymethyl-anthraquinone	Insecticidal
284	*Polygonum punctatum* Ham. Smart weed	Polygonaceae	Wp	Essential oil, gluco., polygenic acid	Antifeedant, insecticidal
285	*Pongamia glabra* (Syn. *Pinnata*) Linn. Pongam oil tree, Indian beech	Leguminosae	Sd, L	Essential fatty acid, Bitter subst. Karanjin	Antifeedant, JHA, insecticidal
286	*Premna integrifolia* Linn Head ache tree	Verbenaceae	L	Alkaloid—preminine	Insecticidal
287	*Prunus armeniaca* Linn Apricot	Rosaceae	Fr, L	Fatty oil, lycopin, ethereal oil	Antifeedant, repellent
288	*Psidium pyriferum* Linn Peru	Myrtaceae	L	Essential oil, eugenol pectin, glycoside	Insecticidal
289	*Psoralea corylifolia* Linn Bakuchi	Leguminosae	Sd	Essential oil, Bakuchiol	JHA, insecticidal

(Contd.)

TABLE 3.1   (CONTINUED)

No.	Botanical and common name	Family	Part used	Biologically active ingredient(s)	Mode of activity
Indigenous plants					
290	*Randia dumetorum* Lamk Emetic nut	Rubiaceae	Fr, Br	Saponin, essential oil, acid resin	Insecticidal
291	*Raphanus sativus* Linn Radish/Karad	Cruciferae	Sd, O	Alkaloid—spirobrassinin	Insecticidal
292	*Rhizophora apiculata* Blume True mangrove	Rhizophoraceae	Br	Tannin	Larvicidal
293	*Rhizophora mucronata* Lam Boriti poles	Rhizophoraceae	Br	Tannin	Larvicidal
294	*Ricinus communis* Linn Castor/Erand	Euphorbiaceae	Sd, L	Alkaloid, ricinine, toxalbumin ricin	Insecticidal
295	*Ruta graveolens* Linn Rue	Rutaceae	L	Glucoside, rutin, essential oil	Repellent
296	*Salvadora persica* Linn Toothbrush tree	Salvadoraceae	Fr, L	Alkaloid, fatty oil and ethereal oil	Insecticidal
297	*Santalum album* Linn Sandalwood	Santalaceae	Wd	Essential oil	Repellent
298	*Sapindus rarak* DC. Soap nut	Sapindaceae	Fr	Saponin, bitter substances	Insecticidal
299	*Sapindus trifoliatus* Linn Three-leaved soapberry	Sapindaceae	Fr	Saponin	Insecticidal
300	*Sarcostemona acidum* R. Moon plant	Asclepiadaceae	St	Bitter substances	Insecticidal

301	*Saussuraa lappa* C.B. Clarke Costus	Asteraceae	Rt	Essential oil, resin, alkaloid, saussurine	Insecticidal
302	*Schkuhria pinnata* Lam Pinnate false thread leaf	Asteraceae	Wp	Schkuhrin I, II	Antifeedant, cytotoxin
303	*Schleichera oleosa* Lour Ceylon oak	Sapindaceae	Br, Sd	Cyanosenete glucoside, fixed oil	Larvicidal
304	*Securinega virosa* BAILL White berry bush	Euphorbiaceae	Wp	Alkaloid, tannin	Insecticidal, repellent
305	*Semecarpus anacardium* Linn F. Marking nut	Anacardiaceae	Fr	Anacardic acid, cardol catechol, bhilawanol	Antifeedant
306	*Sesamum indicum* Linn Sesame	Pedaliaceae	Sd	Flavonoid, glucoside, pedalin, gummy material	Synergists
307	*Sesamum orientalis* Linn Til / Sesame	Pedailiaceae	Sd	Gummy matter	Synergists
308	*Sida acuta* Barm Morning mallow	Malvaceae	L	Alkaloid—ecdysterone	Antifeedant
309	*Siegesbeckia orientallis* Linn Indian weed	Asteraceae	L	Bitter substances, darutin	Antifeedant
310	*Solanum berthaultii* Hawkes	Solanaceae	Wp	β-Farnesene	Pheromone repellent
311	*Solanum laxum* Linn Potato vine	Solanaceae	Wp	Saponin, alkaloid	Insecticidal
312	*Solanum nigrum* Linn Night shade/poison berry	Solanaceae	Fr	Alkaloid—solanine, saponin	Antifeedant, insecticidal

(Contd.)

TABLE 3.1   (CONTINUED)

No.	Botanical and common name	Family	Part used	Biologically active ingredient(s)	Mode of activity
Indigenous plants					
313	*Solanum surattense* Burm F. Thorny night shade	Solanaceae	L	Gluco-alkaloid—solanocarpine	Antifeedant
314	*Solanum trilobatum* Linn Pea egg plant	Solanaceae	Rt, L	Bitter, alkaloid—solanine	Antifeedant
315	*Solanum tuberosum* Linn. Potato	Solanaceae	Tb, Sp	Solanum alkaloid	Insecticidal
316	*Solanum xanthocarpum* Schrad Thorny night shade/Thai egg plant	Solanaceae	Rt, Be, ju	Gluco-alkaloid—solanocarpine	Antifeedant, larvicidal
317	*Sophora mollis* Grah Buna	Leguminosae	Wp	Bitter substances	Insecticidal
318	*Sorghum bicolor* Linn Jowar	Gramineae	St	Glucd. dhurrin, HCN	Antifeedant
319	*Sphaeranthus indicus* Linn Indian sphaeranthus	Asteraceae	Fl, L	Alkaloid, essential oil, terpenoids	Repellent and antifeedant
320	*Spilanthes acmella* Murr Tooth ache plant/para cress	Asteraceae	Fl	Spilanthol	Insecticidal
321	*Suaeda monoica* Forsk Greater Indian saltwort	Chenopodiaceae	Wp	Essential oil, glucoside	Larvicidal
322	*Swertia corymbosa* Wight Nilavembu	Gentianaceae	L	Bitter substance chiratin, ophelic acid	Antifeedant

323	*Swietenia mohagoni* L. West India mahogany	Meliaceae	Wp	Tetranor-triterpenoid, swietenine	Pheromone repellent, antifeedant
324	*Syzygium aromaticum* Linn Clove	Myrtaceae	Fl. Bd	Essential oil eugenol	Larvicidal
325	*Syzygium cumini* (Linn) Skeels Jamun	Myrtaceae	Sd. L	Glucoside, tannin, isoquercitin,myricetin	Insecticidal
326	*Tabernanemontana coronaria* R. Br. Tagar	Apocyanaceae	L	Alkaloid—voacristine, coronaridine.	Insecticidal, ovicidal
327	*Tabernaemontana divaricata* Linn East-Indian Rosebay	Apocynaceae	Br	Tabernae, montanine and coronarine	Insecticidal
328	*Tagetes erecta* Linn Marigold/Zendu	Asteraceae	Fl	Essential oil, pigment quercetugatin	Repellent, insecticidal
329	*Tagetes minuta* L. Wild marigold	Asteraceae	Wp	Bitter substances, essential oil, terepenes	Repellent, insecticidal
330	*Tamarindus indica* Linn Tamarind	Leguminosae	Fr, L	Oxalic acid	Insecticidal, antifeedant
331	*Tanucetum vulgare* L Common tansy	Asteraceae	Wp	Terpinone, thujone, carvone	Feeding deterrent
332	*Tephrosia candida* DC White hoary pea	Leguminosae	Wp	Tephrosine, rotenone, deguelin	Insecticidal
333	*Tephrosia purpurea* Pers Wild indigo	Leguminosae	Sd. L	Rotenone, tephrosine	Feeding deterrent
334	*Tephrosia toxicaria* L. Hoary pea	Leguminosae	Wp	Rotenone, tephrosine	Insecticidal

(*Contd.*)

TABLE 3.1 (CONTINUED)

No.	Botanical and common name	Family	Part used	Biologically active ingredient(s)	Mode of activity
Indigenous plants					
335	*Thevetia neriifolia* Juss Yellow oleander	Apocynaceae	L	Glucoside, thevetin	Insecticidal
336	*Thevetia peruviana* Merrill Yellow oleander	Apocynaceae	L, Ju	Nerifolin, cardiotonic glycoside	Feeding deterrent, chemosterilant
337	*Thuja orientalis* Linn Oriental arbor vitae	Cupressaceae	L	Essential oil, tannin	Antifeedant, larvicidal
338	*Thymus serpylium* Linn Creeping thyme	Labiatae	Sd	Essential oil—cymene, terpene	Growth disruptor
339	*Tinospora cordifolia* Miers Gulbel	Menispermeaceae	L	Berberine, bitter substances	Insecticidal
340	*Tribulus terrestris* Linn Yellow vine/puncture vine	Zygophyllaceae	Fr	Alkaloids, essential oil, resin, nitrate	Antifeedant
341	*Trichilia pallida* Smartz Gaita	Meliaceae	L	Saponin	Insecticidal
342	*Tridax procumbens* Linn Tridax daisy	Asteraceae	Fl	Essential oil	Insecticidal
343	*Trigonella foelium-graecum* Linn Fenugreek	Leguminosae	Sd, L	Alkaloid, saponin, essential oil	Antifeedant, repellent
344	*Vetivera zizanoides* Linn Vetiver	Gramineae	L, Rt	Essential oil, ketone and β-vetivone	Repellent

345	*Vinca rosea* Linn Periwinkle	Apocynaceae	L	Alkaloid, essential oil	Repellent, insecticidal
346	*Vitex negundo* Linn Chaste tree	Verbenaceae	Rt, L	Alkaloid nishindine, essential oil	Insecticidal, repellent, ovocidal
347	*Vitex pubescens* Vahl Hairy-leaved molave	Verbenaceae	L	Alkaloid, resin	Growth disruptor
348	*Vitex trifoliata* Linn Three-leaved chaste tree	Verbenaceae	L	Essential oil, alkaloid	Antifeedant, larvicidal
349	*Warburgia salutaris* (G. Bertol). Chiov. Pepper-bark tree	Cancellaceae	Wp	Warburganal, polygodial, ugandensidial	Antifeedant
350	*Withania somnifera* Dunal Ashwagandha	Solanaceae	L	Alkaloid	Insecticidal
351	*Zanthoxylum alatum* Roxb Winged prickly ash	Rutaceae	Sd, Br	Essential oil, bitter herberine, resin	Larvicidal
352	*Zanthoxylum budrunga* (Roxb.) DC Prickly ash	Rutaceae	Fr	Alkaloid, essential oil, quinazolinone	Larvicidal
353	*Zanthoxylum monophyllum* Lam Yellow prickle	Rutaceae	Rt	Zanthophylline	Feeding deterrent
354	*Zingiber officinale* Rosc Ginger	Zingiberaceae	Rh	Essential oil—gingerol, zingiberine	Repellent, insecticidal

L—Leaves, Rh—Rhizome, Rst—Root stock, Br—Bark, Bd—Bud, Sd—Seed, N—Nectar, Bb—Bulb, Wp—Whole plant, Rt—Root, Fr—Fruit, Fl—Flower, St—Stem, Sh—Shoot, Be—Berry, Wd—Wood, Tb—Tubers, Lx—Latex, Tw—Twig, Ju —Juice, Sp—Sprout, RBr—Root bark, Pd—Pod, Hw—Heartwood, G—Gum, O—Oil, MC—Male cone, SO—Seed oil.

TABLE 3.2   FAMILIES HAVING PESTICIDAL PLANTS

No.	Plant family	Number of plant species		
		Exotic	Indigenous	Total
01	Acanthaceae	–	05	05
02	Amaranthaceae	–	02	02
03	Amaryllidaceae	–	01	01
04	Anacardiaceae	–	03	03
05	Annonaceae	02	04	06
06	Apiaceae	02	06	08
07	Apocynaceae	01	10	11
08	Araceae	01	07	08
09	Aristolochiaceae	–	02	02
10	Ascleptadaceae	01	03	04
11	Asteraceae	08	28	36
12	Berberidaceae	–	01	01
13	Bignoniaceae	–	01	01
14	Buixaceae	–	01	01
15	Boraginaceae	03	–	03
16	Burseraceae	01	02	03
17	Caesalpiniaceae	01	–	01
18	Campanulaceae	–	01	01
19	Canellaceae	–	01	01
20	Cannabinaceae	–	01	01
21	Capparidaceae	01	04	05
22	Caprifoliaceae	01	–	01
23	Caricaceae	–	01	01
24	Caryophyllaceae	02	–	02
25	Casuarinaceae	–	01	01
26	Celastraceae	02	02	04
27	Chenopodiaceae	04	01	05
28	Clusiaceae	–	01	01
29	Convolvulaceae	–	05	05

(Contd.)

TABLE 3.2 (CONTINUED)

No.	Plant family	Number of plant species		
		Exotic	Indigenous	Total
30	Crassulaceae	–	02	02
31	Cruciferae	–	06	06
32	Cucurbitaceae	01	06	07
33	Cupressaceae	–	03	03
34	Cyperaceae	01	02	03
35	Dipterocarpaceae	01	–	01
36	Dioscoreaceae	–	04	04
37	Ericaceae	03	02	05
38	Euphorbiaceae	03	17	20
39	Fagaceae	–	06	06
40	Flacourtiaceae	01	01	02
41	Gentianaceae	–	01	01
42	Ginkgoaceae	–	01	01
43	Gramineae	–	04	04
44	Guttiferae	01	–	01
45	Hippocastanaceae	01	–	01
46	Hippocrateaceae	–	01	01
47	Hypoxidaceae	–	01	01
48	Juglandaceae	01	01	02
49	Labiatae	17	17	34
50	Lauraceae	02	04	06
51	Lecythidaceae	–	02	02
52	Leguminosae	17	31	48
53	Liliaceae	05	04	09
54	Linaceae	–	01	01
55	Loganiaceae	01	–	01
56	Lythraceae	–	01	01
57	Malpighiaceae	–	01	01

(Contd.)

TABLE 3.2 (CONTINUED)

No.	Plant family	Exotic	Indigenous	Total
		Number of plant species		
58	Malvaceae	–	04	04
59	Magnoliaceae	–	01	01
60	Marantaceae	–	01	01
61	Meliaceae	–	10	10
62	Menispermaceae	01	04	05
63	Mimosaceae	–	01	01
64	Moraceae	–	02	02
65	Musaceae	01	02	03
66	Myristicaceae	–	01	01
67	Myrsinaceae	01	02	03
68	Myrtaceae	06	08	14
69	Ochnaceae	–	01	01
70	Oleaceae	01	02	03
71	Palmae	–	01	01
72	Papaveraceae	–	01	01
73	Pedaliaceae	–	02	02
74	Phytolaeaceae	–	01	01
75	Pinaceae	–	04	04
76	Piperaceae	–	05	05
77	Plumbaginaceae	–	01	01
78	Poaceae	01	07	08
79	Polygonaceae	–	03	03
80	Polypodiaceae	01	01	02
81	Pontederiaceae	–	01	01
82	Primulaceae	–	01	01
83	Ranunculaceae	06	05	11
84	Rhizophoraceae	–	02	02
85	Rosaceae	–	01	01

(*Contd.*)

TABLE 3.2 (CONTINUED)

No.	Plant family	Number of plant species		
		Exotic	Indigenous	Total
86	Roxburghiaceae	01	–	01
87	Rubiaceae	–	04	04
88	Rutaceae	07	16	23
89	Salicaceae	01	–	01
90	Salvadoraceae	01	01	02
91	Sapindaceae	03	04	07
92	Sapotaceae	–	04	04
93	Santalaceae	–	01	01
94	Scitamineae	–	01	01
95	Simaroubaceae	02	04	06
96	Solanaceae	07	15	22
97	Tiliaceae	–	01	01
98	Verbenaceae	01	13	14
99	Vitaceae	–	01	01
100	Zingiberaceae	02	03	05
101	Zygophyllaceae	–	01	01
	Total	**128**	**358**	**486**

Arora, R. and Dhaliwal, G.S. (1994). "Botanical pesticides in insect pest management". In: G.S. Dhaliwal and Kansal, B.D. (eds). *Management of Agricultural Pollution in India.* Commonwealth Publishers, New Delhi, India. pp. 213–245.

Caius, J.F. (1998). *The Medicinal and Poisonous Plants of India.* Scientific Publication, Jodhpur, India.

Chopra, R.N., Nayar, S.L. and Chopra, I.C. (1956). *Glossary of Indian Medicinal Plants*. CSIR, New Delhi.

Chopra, R.N. Badhwar, R.L. and Nayar, S. L. (1986). *Insecticidal and Piscicidal Plants of India*. CSIR, New Delhi.

Dhaliwal, G.S. and Arora, R. (2003). Botanical pesticides. In: *Principles of Insect Pest Management*. Kalyani Publishers, New Delhi.

Frank Beye. (1978). Insecticides from the Vegetable Kingdom. In: *Plant Research and Development*. Institute of Pharmaceutical Biology, University of Freiburg. 7: 13–31.

Jacobson Martin and Crosby, D.G. (1971). *Naturally Occurring Insecticides*. Marcel Dekker, Inc., New York. pp. 177–239.

Jacobson, M. (1990). *Glossary of Plant derived Insect Deterrence*. CRC Press, Boca Raton, Florida, USA.

Metcalf, R.L. and Metcalf, E.R. (1992). *Plant Kairomones in Insect Ecology and Control*. Chapman and Hall, New York, USA.

Murugan, K. and Babu, R. (1998). Impact of certain plant products and *Bacillus thuringensis* Berliner subspecies Kurstaki on the growth and feeding physiology *of Helicoverpa armigera* (Hubner). *Sci. Indian. Res.* 57: 757–765.

Satyavati, G.V. Raina, M.K. and Sharma, M. (1976). *Medicinal Plants of India*, Vol. I. ICMR, New Delhi.

Saxena, R.C. and Shrivastava, R.K. (1993). *Insect Pest Management through Natural Products*. Ashish Publishing House, New Delhi. pp. 13–30.

Sharma, R.N. (1982). Development and utilization of plant products for insect control-A comprehensive approach". In: *Cultivation and Utilization of Medicinal Plants*. Atwal, C.K. and Kapur, B.M. (eds.). RRL, CSIR, Jammu-Tawi. pp. 657–667.

Sharma, R.N. and Nagasampagi, B.A. (1979). Workshop on Futurology on use of Chemicals in Agriculture with Particular Reference to Future Trends in Pest Control. Coimbatore.

Singh, R.P. (2000). Botanicals in pest management: An ecological prospective. In: *Pesticides and Environment*. Dhaliwal G.S. and Singh, B. (eds.). Commonwealth Publishers, New Delhi. pp. 279–343.

4

INSECTICIDES OF PLANT ORIGIN—A REVIEW

4.1 INTRODUCTION

The pesticidal properties of many plants have been known for a long time and natural pesticides based on plant extracts such as rotenone, nicotine, pyrethrum, neem, ryania, sabadilla and a number of other lesser botanical pesticides were used to protect agricultural crops from insect and non-insect pests in different parts of the world. For thousands of years, people in India placed neem leaves in their beds, books, grains, bins, cupboards and closets to keep away troublesome insects. However, the efficacy, simplicity, flexibility and economy of DDT and other synthetic insecticides developed during 1940–1950 made the use of botanical pesticides nearly obsolete. Organic pesticides are known for their rapid knock-down effect, but their increscent use has resulted in several environmental as well as biological problems (Verma and Dubey, 1999). Misuse of non-selective chemicals can wipe out the natural enemies and induce problems, with development of resistance. About 450 pest species of insects and mites have developed resistance to one or more synthetic pesticides (Georghiou, 1986).

The success of the pyrethroids has shown the pest control potential of plant-derived substances and has revitalized the interest in plants that contain chemical compounds with pesticidal properties. Jacobson was one researcher who identified the pyrethrins from plants, and covered more than 3000 plant species during the period 1941–1971. More detailed information on chemical structure, isolation and properties of major botanical insecticides such as pyrethroids, rotenoids and tobacco alkaloids was given by Jacobson and Crosby (1971). Graigne *et al.* (1985) have compiled a global database which is divided into three sections:

1. A list of 2400 potential pest control plant species with their active ingredients and mode of action on pest.

2. Different types of pests (nematodes, fungi and bacteria) and plants that control them.

3. Poisonous plants and plants controlling human and animal pests.

The neem tree, *Azadirachta indica*, is so far the most promising example of plants currently used for pest control as well as for many other purposes (NRC, 1992). Since the early 1970s, much research has been carried out on the pesticidal properties of the neem and the results have been published in the Proceedings of the various International Neem Conferences (Schmutterer and Ascher, 1981; 1984; 1987) supported by Deutsche Gesellshaft Fur Technischez Zusammenarbeit (GTZ, 1980), who also have produced manuals for pest control of field crops. A summary of how neem products are used as bio-pesticides, the mode of action, effects on pests and natural enemies was also presented by Schmutterer (1990). Recently, the effects of different neem formulations were tested to control the spotted spider of tomato (Knapp *et al.*, 2003), evaluation of neem seed extract for the control of groundnut leaf spot (Alabi and Olorunju, 2004) and to control maydis leaf blight of maize *in vitro* (Jha *et al.*, 2004).

In general, plants with pesticidal properties can be exploited in three ways. Firstly, by using parts of plants or whole, in powder or as crude extracts in water or other solvents, secondly as purified extracts like rotenone and finally by using a synthesized chemical compound which could be produced industrially. Today there is considerable interest among biochemists and botanists to screen plants for secondary chemical compounds, which could be used for developing medicals and pesticides (Downum, 1993; Dayal *et al.*, 2003).

There is an urgent need to build up reliable food production systems in developing countries, under the paradigms of sustainable agriculture and self-reliance. Different organizations have tried to channel their aid to small farmers in the tropics in order to facilitate their use of resources that are locally available and that can be cheaply maintained. From this perspective natural crop protection could have an important role to play in the many situations where pests seriously hamper agricultural production. The possibility of making raw extracts from plants that are grown in the farmer's neighbourhood assists in ensuring self sufficiency and gives him/ her a cheap alternative to conventional pesticides which often are imported using foreign exchange. Large-scale production of natural pesticides based on crude or purified extracts could also provide an income for people in the countryside.

Efforts have already been made to support the development and use of natural pesticides and many NGOs in particular have shown an interest in this field of bio-pesticides in developing countries (Van Latum, 1991). The indiscriminate and excessive use of organic pesticides has posed a threat to the environment. Emphasis is given to products that are biodegradable and eco-friendly to protect crops or grains during storage against insect infestation (Arthur, 1996). The effectiveness of biopesticides as grain protectants has been studied (Lohra *et al.*, 2000). Recently, the bioactivity of carvone was checked (Tripathi *et al.*, 2003) and the comparative efficacy of some botanicals as protectants against *Sitophilus oryzae* in rice and its palatability (Dayal *et al.*, 2003) was tested. This has once again focused our attention on botanical insecticides and other alternative means of insect control, i.e., today's need. The available data particularly on botanical pesticides or insecticides of plant origin have been dealt with in this chapter, which shall provide a database to conduct research on related aspects of insect pest control. This chapter contains 10 exotic and 11 indigenous plants that have been reported earlier as insecticidal, insect repellent, antifeedant, oviposition and feed deterrent, etc. Such a crop science database may be useful in the developmental strategies and technology for plant protection in the 21st century.

* Strategies in integrated pest management,

* resistance of crop to pest,

* biological, chemical and botanical pesticides and their application,

* biotechnology and molecular biology in plant protection,

* information technology in plant protection and pest prediction

* research in different disciplines of plant protection in various crop systems such as cereal, fibre, cash crop, orchid, forestry and ornamental plants,

* weed science,

* pest management for pre-planting and post-harvesting

* safety for consumers, operators and the environment

Some important issues in the present agricultural system include.

4.2   SOME ESTABLISHED INSECTICIDAL PLANTS

4.2.1 Exotic Plants

i.   *Chrysanthemum cinerariaefolium* Vis

Family	Compositae
Distribution	Yugoslavia, Japan, East Africa and S. America
Part used	White flower heads
Chemical ingredients	Pyrethrum
Importance	It is a daisylike perennial plant of the genus *Chrysanthemum* belonging to the family Compositae. It originated in the Dalmatian

Mountains of Yugoslavia and its cultivation spread throughout the world at the turn of the century. In India *Chrysanthemum cinerariaefolium* and *Chrysanthemum coccineum* are reported to be cultivated. There are about 300 species of herbs and undershrubs of which only a few yield commercial insecticides. The term pyrethrum is commonly applied to the dried flower heads of any of the three species. After the First World War, Japan became the principal exporter. The toxicity of Pyrethrum is considerably increased by the addition of small quantities of rotenone or nicotine (Jung, 1938). Large doses are needed to bring about death (Shepard, 1951). The following contributors reported the insecticidal nature of this plant.

1. Conucher (1980)—pyrethrum is effective against numerous caterpillars, beetles, aphids, mites, locusts, thrips, moths, etc.

2. Gulati, *et al.* (1982)—pyrethrum is one of the safest insecticides known. It has very low mammalian toxicity and is rapidly metabolized if accidentally swallowed.

3. Graigne *et al.* (1985)—the active ingredient obtained from white flowers; it is pure contact poison and has insecticidal, repellent and antifeedant properties. Through its effect as a nerve poison, it produces erratic movement, excitement and finally paralysis (knockdown effect).

The insect can, however, recover from the effect within 24 hours.

4. Bowers *et al.* (1987) isolated a polyacetylenic sulphoxide compound having anti-juvenile hormone activity against *Oncopeltus fasciatus*, from *C. coronarium*. This study is the first to reveal an antihormonal activity for the polyacetylenes. The essential oil of *C. balsamita* has been tested in the laboratory against the aphid, *Metopolophium dirhodum*, on wheat seedlings in pots.

5. Bestmann *et al.* (1987) tested the essential oil of *C. balsamita* in the laboratory against the aphid, *Metopolophium dirhodum*, on wheat seedlings in pots. It revealed insecticidal properties that were attributed to the presence of Pyrethrin-I in oil and appeared to depend on the date on which the *Chrysanthemum* plants were harvested. They further studied that oil from the leaves was more effective when these were gathered in June than in October. Secondly, leaf oil was more effective than oil obtained form flower heads.

6. Rajasekeran *et al.* (1996) studied the repellent and knockdown properties of pyrethrins obtained from *Chrysanthemum cinerariaefolium* (Figure 4.1) and compared it with natural-derived standard against 3-day-old female *Culex quinquefasciatus*.

The knockdown insects did not recover within 24 hours. Extract applied at 1 ppm to wheat grains exhibited similar repellency of 2–3 week-old *Tribolium castaneum*.

FIGURE 4.1 *CHRYSANTHEMUM CINERARIAEFOLIUM* L.(*SEE* PLATE 1.1)

7. Perez and Pascaul (1999) studied the anti-insect, repellence and volatile toxicity of blue essential oil obtained from flower heads of *C. coronarium* L. against white fly *Bemisia tabaei* and the stored product pests *Tribolium castaneum, Acanthoscelides obtectus* and *Ephestia kuchniella*. Topical application gave high mortality with a fast knockdown effect in these insect pests. The experimental results show the potentiality of in using this natural product in pest control.

8. Meshram (2000) tested crude extract of fresh leaves of *Chrysanthemum indicum*, against 3rd instar larvae of *Dalbergia sissoo*

defoliator and *Plecoptera reflexa*. They report antifeedant and insecticidal activities.

ii.   *Derris elliptica* (Roxb) Benth. (Derris)

Family	Leguminosae
Distribution	Tropical rainforests of the Malay Archipelago
Part used	Root
Chemical ingredients	Rotenone
Importance	Derris is a small shrub originally seen in the tropical rainforests of the Malay Archipelago. It grows in lowland areas and does not thrive at higher altitudes. Its roots contain the active substance, rotenone. When grown in the shade, *Derris malaccensis* requires a minimum period of 1½ years to produce worthwhile rotenone, but when grown in full sunlight, it needs only nine months for the roots to develop sufficiently. Derris thrives on many soils but particularly on loams and clays. The active ingredient rotenone acts as a contact and stomach poison. It has very good insecticidal and repellent property. It is also effective against the fungus, *Pyricularia oryzae*. Derris is most effective against larvae in the young stages. Caterpillars, aphids and beetles are very vulnerable, but it is not effective against cockroaches. Derris is affected by sunlight, oxygen and temperature. After a week in strong sunlight, derris dust is no longer effective. In the shade, it retains its properties for a fortnight. Derris is harmless to bees, but toxic to fish (Anonymous, 1928).

FIGURE 4.2   *DERRIS ELLIPTICA* (*SEE* PLATE 1.3 A&B)

1. Visetson *et al.* (2001) evaluated ethanol extract of this plant (containing, the rotenone) and used it as a larvicide against 3rd instar larvae of the diamondback moth. They showed significantly lowered number of larvae in the field. Derris extract induced 10–20% of all enzyme activity.

2. Unjitwatana *et al.* (2003) studied the acute toxicity of derris extract and its formulations in the fish, *Tilapia nilotica*. After 96 hours, the LC_{50} value of rotenone from derris extracts and derris formulations were 0.0007 and 0.0008 mg/litre respectively. These small concentrations are enough to inhibit brain cholinesterase activity.

iii. *Haplophyton cimicidum*
    (Mexican cockroach plant)

Family	Apocynaceae
Distribution	Central America, Mexico and Guatemala
Part used	Leaves, stems

Chemical ingredients Haplophytine, cimicidine and other alkaloids.

Importance It appears to have found insecticidal use in central Mexico and Guatemala. The insecticidal nature of this plant has been reported by the following researchers.

FIGURE 4.3 *HAPLOPHYTON CIMICIDUM* (SEE PLATE 1.2)

1. Plummer (1938) stated that "cockroach powder" is prepared from the dried leaves and appears to be a contact and stomach poison, and has insecticidal activity.

2. Heal *et al.* (1950) reported that laboratory tests showed a high level of toxicity to the German cockroach, milk weed bug, clothes moth, black carpet beetle and larvae of the mosquito.

3. Rogers *et al.* (1952) reported that the powder mixture was effective against

varieties of pests like Mexican fruit flies, European corn borer, Mexican bean beetle larvae, Colorado potato beetle, grasshopper, egg-plant lace bug and codling moth.

4. Iwuala (1981) reported that sprays containing 3.3% of leaf powder were effective against flying insects.

iv. *Mammea americana* L. (Mammey)

Family	Guttiferaceae
Distribution	Caribbean and in Northern South America, Africa and Asia
Part used	Ripe seeds, leaves and bark
Chemical ingredients	Mammein
Importance	This tree grows to a height of 20 m. It thrives up to 1000 m and there are two harvests a year and each tree yields 300–400 fruits.

1. Plank (1950) reported that all major parts of the plant were found to be insecticidal. Powdered mammey seeds or their extracts act as a contact insecticide and probably have stomach action also. A variety of the common insect pests of plants are controlled and insecticidal preparations also received wide use against body pests of man and animals. After 4 days of sun, wind and dew, mammey powder was still an effective poison. In the West Indies the resin of the mammey apple tree is used against flies.

FIGURE 4.4   *MAMMEA AMERICANA* L. (*SEE* PLATE 2.2)

2. Greenspan *et al.* (1996) showed that the extract of seeds prepared with kerosene can be used effectively against household vermin. They also reported that the seeds and leaves extract have a historical use as a biopesticide. The insecticidal ingredients are active against *Blattella germinica*, *Periplaneta americana* and *Plutella xylostella*. These materials represent renewable source of bioinsecticides for agriculture and should regrade interest in coumarin-type compounds for novel pesticidal action.

v.   *Pachyrrhizus erosus* (Yam Bean)

Family	Leguminosae
Distribution	South America, Mexico, Indonesia and West Indies
Part used	Ripe seeds
Chemical ingredients	Pachyrrhizin, erosmin, pachyrrhizone, dolineone

Importance

Yam bean is a leguminous shrub. It is indigenous to South America and is cultivated as a crop in Mexico. The plant is also well known in the orient and Indonesia. Its Mexican name is "Jicama" and it is called "Sinacamas" in the West Indies. Yam bean seed powder is very toxic to fish as might be expected from their rotenoids and furocoumarin constituents. Although the insect powder apparently has received long use against the body pests of humans, there is evidence that its toxicity to mammals also may be significant.

FIGURE 4.5   *PACHYRRHIZUS EROSUS (SEE* PLATE 1.4 A&B)

1. Boorsma (1910) reported that the large seeds have received continued local use as an insecticide in both hemispheres, although other plant parts also have been claimed to show mild activity. The dry, powdered seed may be used directly in the form of a decoction in water.

2. Hansberry and Lee (1943) reported that the seed powder is insecticidal both upon contact and injection, and its extracts often are more active than those of total plant extract.

3. Ghatak and Bhushan (1995) evaluated the ovicidal activity on eggs of *Corcyra cephalonica*. They showed that there was no egg hatching at 6 days after treatment with 2% petroleum ether extracts.

4. Alavez-Solano *et al.* (1998) screened the solvent extract of seeds of this plant for insecticidal properties. The seed contained the rotenone erosone, paquirrizone, dolineone, paquirrizine and isoflavone dehydroneotenone. These exhibited antifeedant and insecticidal properties against *Sitophilus zeamais*.

vi. *Quassia amara* L. (Quassia Wood)

Family	Simaroubaceae
Distribution	Central America, Brazil and Surinam.
Part used	Roots, leaves, bark and wood.
Chemical ingredients	Amaroids, quassin and neoquassin.
Importance	It is a small, 4–6 m tall, tropical tree which is distributed in Central America, Brazil and Surinam (Dutch Guiana), and so is often referred to as "Surinam quassia". A related West Indies shrub, *Aeschrion excelsa* ("Jamaica quassia"), eventually is the principal source of the insecticide.

1. Busbey (1939) stated that it appears very likely that even before the turn of the 19th century, aqueous extracts were in use against insects.

2. Roark (1947) showed that due to its water-soluble ingredients, quassia sprays are used against sap feeders when taken up by the roots. Quassia works systematically and is transported into the leaves where it acts as a stomach poison.

3. Atwal and Panji (1964) reported that the plant acts as a contact, systemic and stomach poison. It has insecticidal, larvicidal and nematicidal properties.

4. Iwuala (1981) reported that the entire plant contains insecticidal properties, but the roots, leaves and bark all contain quassin to a small degree. In India, farmers use *Pricasma excelsa*, which is either closely related to or identical with *Aeschrion excelsa*. Beneficial insects like ladybirds and honeybees are not killed by quassia spray preparation.

5. Bloszyk *et al.* (1995) evaluated antifeedant activity of African traditional medicinal plant, *Quassia africana,* against adults and larvae of stored grain pests viz. *Tribolium confusum*, adults of *Sitophilus granarius* and larvae of *Trogoderma granarium*. This plant extract exhibited good antifeedant activities.

6. Mambelli *et al.* (1996) extracted active components from the wood of *Quassia amara*, containing high content of active principles quassin and neoquassin. These are good botanical insecticides and have been screened for activity against insect pests in the field.

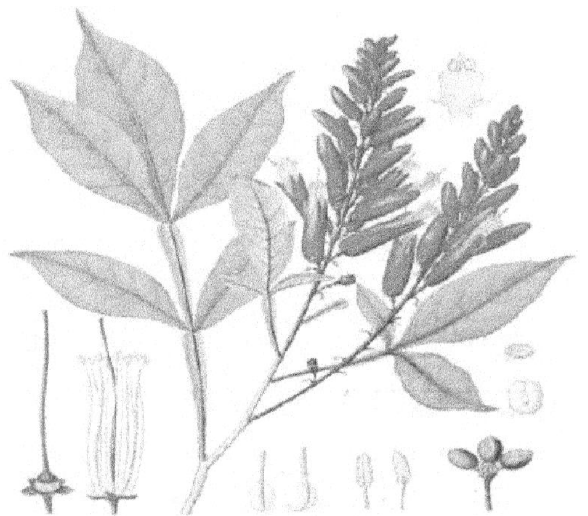

FIGURE 4.6 *QUASSIA AMARA* L. (*SEE* PLATE 2.1A&B)

7. Latif *et al.* (2000) studied aerial parts of this plant for their phytochemical content and insecticidal activity. They purified the active insecticidal compound, quassinoid. The lethal activity of methanol extract was found against *Tetranychus urticae*, *Myzus persicae* and *Meloidogyne incognita* at 10,000 ppm.

8. Manecebo *et al.* (2000) carried out laboratory bioassay for the antifeedant activity of wood and leaf extract on mahogany shoot borer, *Hypsipyla grandella* larvae at variable doses and showed

antifeedant activity in all extracts (0.1, 0.316, 1.0, 3.16 and 10%), the highest activity seen with high concentrations.

vii. *Ryania speciosa* Vahl. (Ryania)

Family	Flacourtiaceae
Distribution	Northern part of S. America and Amazon basin.
Part used	Root and stalk.
Chemical ingredients	Ryanodine and its derivatives
Importance	Ryania is a unique example of a commercially successful natural insecticide. The active substances ryanodine and its derivatives are more stable than those found in the pyrethrin- and rotenone-bearing plants. Although the major use of Ryania probably was against the

FIGURE 4.7   *RYANIA SPECIOSA* (*SEE* PLATE 2.3)

European corn borer, *Pyrausta nubialis*, the most notable effect is the direct and irreversible contractile action on many vertebrate and invertebrate muscles, although the mammalian uterus was insensitive to this agent. Other effects included a decline of contractile force in cardiac muscle, ATP-dependent uptake of calcium ions and the characteristically high oxygen demand in treated insects just prior to paralysis (Shepard, 1951). It acts as a contact and stomach poison. Its action is slow, but it is highly effective, even when the insect does not appear to have been immediately affected. Eating, movement and breeding activities gradually cease after contact (Iwuala, 1981).

viii. *Schoenocaulon officinale* Grey (Sabadilla)

Family	Liliaceae
Distribution	Venezuela, Colombo and Mexico
Part used	Spikes with immature capsule of seeds
Chemical ingredients	Alkaloids like cevacine, cevadine, veratridine and ceverathrum
Importance	Sabadilla is a perennial, 50 cm tall, lily-like plant. Most insecticidal properties are mainly found in spike with immature capsule of seeds.

1. Allen *et al.* (1944) showed that a preparation of seed extract in kerosene at high temperature (150°C) is effective against a number of agricultural pests and its toxicity is increased when heated without the addition of other substances.

2. Roark (1947) reported that good results were obtained with aqueous extract of Sabadilla prepared with soda-ash against

grasshopper and other insects. Sabadilla preparations are extremely toxic to honeybees. Therefore, care should be taken before its application in field (Figure 4.8).

3. Graigne *et al.* (1985) reported that it shows insecticidal, repellent and rodenticidal properties.

4. Ujvary and Casida (1997) synthesized 3-O- Vanilloylveracevine, an insecticidal alkaloid that is a very potent insecticide against *Musca domestica*, *Periplaneta americana* and *Phaedon cochleariae*.

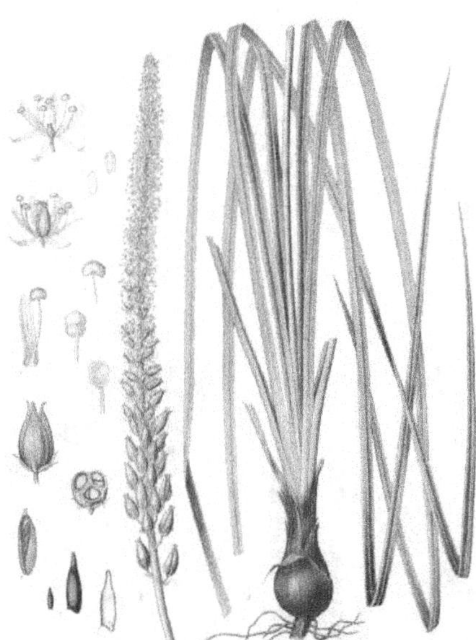

FIGURE 4.8   *SCHOENOCAULON OFFICINALE (SEE* PLATE 3.2)

5. Osario (2004) described the use of this plant product in plant protection, and their advantages over the use of pest-resistant transgenic crops. It is a very good insecticidal plant.

ix. *Tephrosia vogelii* Hook F. (Vogel Tephrosia)

Family	Leguminosae
Distribution	USA
Part used	Root bark
Chemical ingredients	Rotenone
Importance	It is a potential source of rotenone, an important nonresidual insecticide, and also a material used in killing undesirable fish (Blommaert, 1950). It is maintained as a semi-cultivated plant in dooryards in some primitive areas where it is used for poisoning of fish. *Tephrosia* is a short lived, slow-growing, herbaceous, frost-susceptible perennial. Barnes and Freyre (1969) suggested that for commercial production of rotenone, which is derived largely from the leaves, the plants should be grown at the rate of 30,000 to 37,000 per hectare.

FIGURE 4.9   *TEPHROSIA VOGELII* F. (*SEE* PLATE 2.4A&B)

1. Gaskins *et al.* (1972) reported it is a good source of rotenone and can be used to kill insects and fish.

2. Zhang *et al.* (2000) evaluated the antifeedant properties of acetone extracts of this plant in both choice and no-choice bioassay in 4 lepidopteran species, *Pieris rapae, Spodoptera litura, Mythimna separata* and *Plutella xylostella*. In both these methods the plant extracts significantly reduced the feeding behaviour of lepidopteran larvae.

3. Minja *et al.* (2002) conducted a field experiment to evaluate the effect of crude extracts of plant on the incidence of seed damage by insect pests mainly *Helicoverpa armigera, Maruca vitrata, Etiella zinckenella* and *Lampides* sp. It was showed significant reduction in seed damage at early and late pod-formation stages and at pod maturity. Seed weight increased due to insecticide application of leaves extract.

x. *Tripterygium wilfordii* Hook F. (Thunder god vine)

Family	Celastraceae
Distribution	China, North America
Part used	Root bark
Chemical ingredients	Alkaloids—wilfordine, wilforgine, etc.
Importance	*Tripterygium wilfordii* has been employed widely in China as a garden insecticide known as "lei-kung-teng". Aqueous extracts had a direct

effect on the heart, smooth and striated muscles rather than primarily on the nervous system (Figure 4.10).

1. Ta-wang, (1941) reported that although some species of chewing insects are repelled by bark, the stomach-poison effect is quite selective.

2. Lee and Hansberry (1943) found that lepidopteran larvae feeding on treated plant were violently affected and often shrunk to less than one-third their original size. However, the flaccid paralysis was fairly rapid, and the insects frequently could recover when removed to untreated food.

FIGURE 4.10   *TRIPTERYGIUM WILFORDII* HOOK F. (*SEE* PLATE 3.1A&B)

3. Cheng (1945) stated its insecticidal properties against aphids and also claimed that it is safe for man and domestic animals.

4. Beroza, (1953) reported that the root bark extract shows insecticidal property. It contains at least five insecticidal ester alkaloids.

5. Luo *et al.* (2004) reported for the first time the insecticidal activity of 3 active compounds from ethanolic extract of the root bark of this plant. They exhibit insecticidal activity against larvae of *Mythimna separata*. These compounds are triptolide, triptonide and euonine. They also showed antifeedant activity of 3rd instar larvae of *Mythimna separata*.

4.2.2 Indigenous Plants

i.   *Acorus calamus* Linn. (Sweet flag/Vekhand)

Family	Araceae
Distribution	Worldwide
Part used	Rhizome
Chemical ingredients	Essential oil—Azarone (hexone), glucoside, acorine.
Importance	It is a perennial plant native to India which has worldwide distribution because of its high medical value. In India over 4000 kg of dried rhizome per hectare can be harvested annually (Subramaniam, 1948/49). It was reported to be the best plant for control of insect pests by earlier authors.

1. Mathur and Saxena (1975) reported that it acts as an insecticide, and possesses repellent, antifeedant and antifertile properties.

2. Graigne *et al.* (1985) have reported that this plant acts in a variety of ways in various insect pests, especially as repellent and antifeedant against stored grain pests viz. *Callosobruchus chinensis, Sitophilus oryzae* and others.

3. Yadava (1971) showed that when used on crops in store, the germination power is not harmed and the food quality does not suffer.

4. Hiremath and Ahn (1997) obtained methanol extract of this plant by rotary vacuum evaporator at 40°C. The extract was tested against a rice pest, paddy brown hopper (*Nilaparvata lugens*). During tests 30% mortality of adult female hopper was achieved.

FIGURE 4.11   *ACORUS CALAMUS* LINN. (*SEE* PLATE 3.3)

5. Harish Chander *et al.* (1998) reported the effect of rhizome powder on *Tribolium castaneum* (Herbst). The powder was mixed with milled rice at 0.5% (w/w) and stored in jute bags under warehouse condition for 6 months. The powder was more effective in reducing reproduction, and suppression of more than 70% progeny production was recorded.

6. Punam Kumari *et al.* (1999) carried a laboratory experiment to assess the effect of leaf powder of this plant at 0.5, 1.0, 1.5 and 2.0 g doses against *Callosobruchus chinensis*. A high dose (1.5 and 2.0 g) resulted in higher mortality than lower doses of short duration. Instant mortality was observed at any dose.

7. Park, A. C. (2003) evaluated the rhizome from *Acorus gramineus* Soland for its potential to manage stored-product insects against the rice weevil, *Sitophilus oryzae* (Linn), the adzuki bean weevil, *Callosobruchus chinensis* (Linn) and cigarette beetle, *Lasioderma serricorne* (F), which are the most widespread and destructive primary insect pests of stored cereals, stored legumes and stored tobacco and tobacco products respectively. Direct contact and fumigation methods were used and it was observed that this method is very effective in controlling the above pests.

ii.   *Allium sativum* L. (Garlic)

Family	Liliaceae
Distribution	Europe, Central Asia
Part used	Bulb
Chemical ingredients	Volatile oil, essential oil–allyl propyl disulphide. allisatin-I and II
Importance	Garlic is a cosmopolitan plant which grows in temperate zones as well as in the tropics and subtropics. It probably originated in Central Asia from where it spread to the Mediterranean where it still finds its greatest use in the kitchen. It is much cultivated and easy to grow in the field, garden or backyard. For use as an insecticide, it should not be grown with mineral fertilizers since it has been established that heavy doses of fertilizer reduce the concentration of effective substances.

1.   Greenstock (1970) reported that garlic showed insecticidal, repellent, antifeedant, bactericidal, fungicidal, and nematicidal action and was effective against ticks.

2.   Bhatnagar *et al.* (1974) stated that besides its food value, it is one of the important vegetable insecticides used to control insect pest population in the form of aqueous extract or combined with other vegetables like onion, lemon grass, etc.

3.   Gu- YanFang (1997) extracted the oil with petroleum ether to test their repellent action on adult *Tribolium castaneum*.

The oil of this plant was the most repellent at 0.2 ml/dish. Per cent repellency after 72 h exposure was 81 per cent.

FIGURE 4.12   *ALLIUM SATIVUM* L. (*SEE* PLATE 3.4)

4. Ho *et al.* (1997) tested the garlic oil for toxicity against the eggs, larvae and adults of *Tribolium castaneum* and adults of *Sitophilus zeamais*. *T. castaneum* egg mortality increased with garlic oil concentration, complete kill of eggs being achieved at 4.4 mg /cm^2, using the filter paper bioassay. The eggs were the most susceptible stage, followed by adults, 10-day-old larvae and older larvae. *Tribolium castaneum* adults were more susceptible to oil than *Sitophilus zeamais* adults, with LD$_{50}$ values of 1.32 mg/cm^2 and 7.65 mg/cm^2, respectively.

5. Asolkar *et al.* (2000) reported its insecticidal, antibacterial, anti-fungal and nematicidal property.

6. Choudhari *et al.* (2001) reported that the garlic extract provided satisfactory control of sucking pests viz. white fly and aphid, as compared to pyrethroids and endosulfan. They further found that this extract also increases the yield of cottonseeds.

7. Singh and Reena (2003) revealed the bio-efficacy of garlic products like garlic oil, extracts of water, acetone, ethanol, methanol, petroleum ether, benzene, chloroform, etc. which were demonstrated to act as oviposition deterrents, antifeedants, setting inhibitants, larvicides and growth regulators against insect pests.

iii. *Annona squamosa* Linn. (Sitaphal/Custard apple)

Family	Annonaceae
Distribution	India, South-east Asia.
Part used	Unripe fruit, seeds, leaves and roots.
Chemical ingredients	Amorphous alkaloid, resin, hydrocyanic acid, anonaine, essential oil and aporphine alkaloid.
Importance	More than 800 species of small tree and shrubs are known in the Annonaceae family. They do not require special condition of soil or water, but thrive best in places where there is a clear division between the rainy and the dry season

and generally prefer dry sites on which to grow. Many authors have reported its insecticidal nature.

1. Harpar *et al.* (1947) have claimed that *Annona* extract acts as both contact and stomach poison. As a contact poison, the effect was variable.

FIGURE 4.13   *ANNONA SQUAMOSA* LINN. (*SEE* PLATE 3.5)

2. Schmutterer and Ascher (1987) revealed that the most effective application appeared to be against various aphids. Powdered seeds applied to wheat and rice grains act as a protectant against *Sitophilus oryzae* (Linnaeus) and *Callosobruchus chinensis*.

3. Kumar and Thakur (1988) tested petroleum ether extract of seed up to 1000 ppm for antifeedant activity against

4th instar larvae of the noctuid, *Spodoptera litura*. They reported negligeble damage, by larvae, of the treated area of plant leaves.

4. Jacobson (1990) isolated the compound, anonaine, which has been found effective against the infestation by termites, root grubs, etc. whereas recently isolated compounds, annonacin, and annonidines, may prove to be biologically active.

5. Chitra *et al.* (1992) tested petroleum ether extract at 0.5 and 1.0% against 1st and 3rd instar larvae of *Henosepilachna vigintioctopunctata* (Brinjal spotted leaf beetle). The extract gave more than 90% protection to plant and also reduces more than 20% weight of larvae.

6. Arora and Dhaliwal (1994) showed that when seed powder was mixed with other botanical insecticides, they showed disruption of growth, reduced oviposition, reduced adult emergence and moderated toxicity in different species.

7. Ghatak and Bhushan (1995) evaluated in the laboratory the ovicidal activity of crude extract on the egg of *Spilosoma obliqua*. The result revealed that none of the eggs could hatch on treatment with 2% methanol extract.

8. Prasad *et al.* (1996) made an *in vitro* study of seed powder suspension of this plant on *Culex* larvae at 0.025 to 1.5% dose

concentrations. There was no mortality during the 1st hour, whereas from 2 to 12 h, mortality ranged from 2 to 100%. They showed that the rate of mortality increased with increase in concentration of seed extract.

9. Vyas *et al.* (1999) evaluated methanolic seed extract at various doses against 3 important noctuid pests (*Earias vittella* F., *Helicoverpa armigera* Herb and *Spodoptera litura* F.). The data on percentage mortality were 89%, 100% and 44% respectively. In addition to larval mortality, the extracts also reduced larval growth and development, prolonged larval duration to reach pupation and lowered pupal weight resulting in the formation of deformed individuals. Results of the study clearly suggest that this plant (custard apple) can supplement neem formulation in an environmentally friendly pest control programme.

10. Maheshwari *et al.* (2001) tested the bioefficacy of partially purified flavonoids from the foliar extract of this plant, for antimicrobial and insecticidal activity against the common infestants of pulses and the stored grain pest, *Callosobruchus chinensis*. The isolated flavonoids showed very excellent antimicrobial as well as 80% insecticidal activity at a concentration of 0.07 mg/ml.

11. Al-Lawati *et al.* (2002) tested the leaf and seed extracts for insecticidal and repellent properties against the pulse beetle, *Callosobruchus chinensis*. The seed extract recorded 100% mortality within 20–24 h of their exposure to methanol and ethanol extracts respectively.

iv. *Azadirachta indica* A. Juss. (Neem)

Family	Meliaceae
Distribution	Central America, Mexico, most of the Guatemala and Asian countries
Part used	Leaves, stems
Chemical ingredients	Haplophytine, cimicidine and other alkaloids.
Importance	Neem tree is indigenous to India from where it has spread to many Asian and African countries. For centuries, the tree has been held in esteem by Indian folk because of its medicinal and insecticidal value. Due to its legendary insect repellent and medicinal properties, it has been identified as the most promising of all plants by the National Research Council, Washington, USA (NRC, 1992). Neem has assumed the status of an international tree which is evident from the fact that it has been a subject of discussion at six global conferences, viz; Rottach-Egern, Germany (1980), Rauischolzhausen, Germany (1983), Nairobi, Kenya (1986), Banglore, India (1993), Queensland, Australia (1996) and Vancouver, Canada (1999). It was also

discussed in the International Plant Protection Congress at Beijing, China (2004).

All parts of the neem tree possess insecticidal activity but seed kernel is the most active. Neem products exhibit almost every conceivable type of activity against insects. Till today, 450 to 500 species of insects have been tested with neem products and 413 of these are reportedly susceptible at different concentrations (Schmutterer and Singh, 1995). It has been found to possess several types of chemicals, many of them still unknown, which could be exploited for pest management depending upon the nature of the pest. It has been reported to be antifeedant, attractant, repellent, insecticide, nematicide, growth disrupter and antimicrobial (Jacobson, 1975; Attri, 1982). Therefore, commercialization of neem-based pesticides is expanding at a rapid rate. More than three dozen products are marketed in India (Parmar and Ketkar, 1993). Huge literature is available on neem.

1. Ketkar and Schmutterer (1987) evaluated the non-edible oil of neem against the bruchid stored product pest *Callosobruchus maculatus* on cowpea and *Callosobruchus chinensis* on green gram. The neem oil was the best surface protectant giving complete protection at 0.75% and 1.0% even after 150 days. It also reduces growth index and has greatest ovicidal effect on *Callosobruchus maculatus*.

FIGURE 4.14   *AZADIRACHTA INDICA* A.JUSS. (*SEE* PLATE 4.1)

2. Yamasaki *et al.* (1988) reported that salannin ($C_{34}H_{44}O_9$) obtained from neem seeds is a triterpenoid with insect antifeedant and insect growth regulator.

3. Singh and Singh (1996) used different neem formulations and studied repellent, oviposition-deterrent, antifeedant, feeding and contact activities of various agricultural and stored grain pests like desert locust (*Schistocera gregaria*), migratory locust (*Locusta migratoria*), maize stem borer (*Chilo partellus*), cotton boll worms such as pink boll worm (*Pectinophora gossypiella* and *Earias insulana*), (Mango hopper (*Idioscopus clypealis*), green leaf hopper (*Nephotettix* Spp.), tobacco caterpillar (*Spodoptera litura*), diamondback moth (*Plutella xylostella*), gram pod borer

(*Heliothis armigera*), spotted pod borer (*Maruca testulalis*), pod fly (*Melanagromyza obtusa*), pulse beetles (*Callosobruchus maculatus* and *Callosobruchus chinensis*), lesser grain borer (*Rhyzopertha dominica*), rice weevil (*Sitophilus oryzae*), Khapra beetle (*Trogoderma granarium*), and Angoumois moth (*Sitotroga cerealella*).

4. Singh *et al.* (1996) conducted laboratory studies to determine the effect of neem extract on lesser grain borer, *Rhyzopertha dominica* (F), on wheat. 2 and 3% doses were most effective in terms of adult mortality and reduced grain damage.

5. Jeyabalan and Murugan (1997) reported the limnoid compounds deacetylnimbin, 17-hydroxyazadiradione, gedunin, salannin and deacetylgedunin, which are constituents of neem tree show their effects on development, feeding and reproduction in a polyphagus insect *Helicoverpa armigera*. The neem limnoid significantly inhibited growth, longevity and fecundity when administered orally through cotton leaves.

6. Meshram (2000) tested crude extract of the fresh leaves against 3rd instar larvae of the *Dalbergia sissooo* defoliater and *Plecoptera reflexa* Gue. The investigation revealed that the plant has some antifeedant and insecticidal activity.

7. Gupta *et al.* (2001) reported the efficacy of seven neem-based formulations in controlling mustard aphid, *Lipaphis erysimi* Kalt.

8. Karbasayya, Rahman, M.F. (2001) reported the efficacy of neem products in seed treatment for the management of *Meloidogyne incognita* on black gram.

9. Mallapur *et al.* (2001) reported that extracts of neem seed kernel and neem cake effectively controlled the safflower aphid. This botanical pesticide exhibited safflower aphid control comparable to that obtained with synthetic insecticides. As these products are ecologically and environmentally safe and possess high benefit–cost ratio, they can be utilized in the management of safflower aphid.

10. Vijayalakshmi *et al.* (2001) reported the seed coating with neem formulations for management of *Heterodera cajani* and other phytoparasitic nematodes. They showed that use of different neem formulations reduces the population of harmful nematodes in soil in the field, without disturbing the saprophytic nematodes, and increased grain yield.

11. Devaraj and Nandihalli (2002) reported highest pupal mortality (63.35%) of *Helicoverpa armigera* (Hunber) and lowest adult emergence (36.67%) by neem seed kernel.

12. Kalita *et al.* (2002) evaluated the effectiveness of neem powder in the control of bruchid (*Callosobruchus chinensis* L) in stored pulses (grain legumes). The result indicated that neem powder is most effective in controlling pulse bruchids, acts as strong oviposition-deterrent and grain protectant against pulse beetle in storage.

13. Yadav and Bhargava (2002) conducted laboratory studies to determine the effect of neem extract on the stored product pest *Corcyra cephalonica*. Neem extract at 1 ml/100 g seed resulted in the longest total life cycle (57.8 days), highest reduction in adult emergence (85.7%), lowest number of eggs laid per female, highest reduction in egg viability (65.3%) and shortest longevity for males (3.3 days) and females (4.8 days).

14. Knapp *et al.* (2003) prepared different neem formulations, and found that these formulations of neem products cause repellence and oviposition deterrence on the two-spotted spider mite, *Tetranychus urticae* Koch. on tomato (*Lycopersicon esculentum* Mill).

15. Dayal *et al.* (2003) reported the comparative efficacy of some botanicals as protectant against *Sitophilus oryzae* in rice and its palatability. Neem seed oil gives 73.5% protection after 10 days of application.

16. Patil and Goud (2003), evaluated the plant extract for its ovipositional repellent property against *Plutella xylostella* under laboratory conditions. Among the plant products tested, 0.5% neem recorded maximum reduction in egg laying both under no-choice (50.33%) and free-choice (62.43%) condition at 24 hour.

17. Rao *et al.* (2003) reported the potential of plant extract with extract of sweat flag (*Acorus calamus*) and pungam (*Pongamia glabra*) at 1 : 1 : 1 (NSP-I), 2 : 1 : 1 (NSP-II) and 3 : 1 : 1 (NSP- III) (v/v) ratios for the control of *Earias vittella* (Fab). The mixture was more effective than individual treatments. The mortality of shoot and fruit borer were maximum (93.33%) in NSP-I compared to neem alone.

18. Alabi and Olorunju (2004) evaluated neem seed extract along with black soap and cow dung for the control of groundnut leaf spot at Samaru (Nigeria). There was significant high correlation between disease severity score, and pod and haulm yields. Plants sprayed with neem seed extract gave yields higher than the plants that received other treatments, apart from untreated control.

19. Jha *et al.* (2004) reported that neem at 0.2% combination with garlic (0.5%) and onion (0.1%) showed inhibition of fungal sporulation (*Helminthosporium maydis*).

20. Nandgopal and Ghewande (2004) studied use of neem products such as aqueous extracts of neem leaf, bark and seed, neem oil and neem-based formulations for groundnut pest management. They found that the above neem products are highly effective in controlling the population of jassid (*Batclutha hortensis* Lindh), aphids (*Aphis craccivora* Koch), leaf minor (*Aproaerema modicella* Dev.), tobacco caterpillar (*Spodoptera litura* F.) and thrips (*Caliothrips indicus* K.), which are pests of groundnut.

21. Gupta (2005) showed the efficacy of neem leaf and seed kernel extracts mixed with cow urine as an insecticide against mustard aphid, *Lipaphis erysimi* Kalt. It was observed that the treatment significantly reduced incidence of mustard aphid, increased the grain yield of mustard and increased the number as well as activity of the predator coccinellid beetles (*Coccinella septumpunctata, Coccinella transversalis, Cheilomenes sexmaculata*).

v. *Capsicum frutescens* L. (Chilli/Mirachi)

Family	Solanaceae
Distribution	Tropical (South) America and West Indies
Part used	Skin, seeds
Chemical ingredients	Capsaicin

Importance

Chillies are widely distributed in the tropics and subtropics and originated in South America where they were cultivated early. They were introduced to the rest of the world by the Spaniards and the Portuguese. Chilli is occasionally cultivated for fruits. The ripe fruit has insecticidal properties and the effective substances are highest in the skin and seeds. This statement has been proved by the following contributors.

FIGURE 4.15   *CAPSICUM FRUTISCENS* L. (*SEE* PLATE 4.3A&B)

1. Spickett (1955) reported that chilli mixture an often prepared from maximum part of chilli along with different insecticidal plants, especially garlic and onion. This mixture is very effective against insect pests like aphids and 'worst pests' of rice.

2. McKeen (1956) reported that it can also be used against viral infection on plant.

3. In an anonymous report (1977) it was shown that in another mixture prepared using chilli and tubli roots (*Croton tiglium*), dried tobacco leaves were added and used to control vegatable pests.

4. Deb-Kirtanya (1980) showed its insecticidal activity against ants, aphids, caterpillars, Colorado beetle, imported cabbage worm, rice weevil and house pests.

5. Graigne *et al.* (1985) showed that it acts as insect repellent, antifeedant and fumigant, and has virulent activity.

6. EI-Lakwah *et al.* (1996) carried out laboratory experiments on the relative toxicity of acetone and petroleum ether extracts of fruits of *Capsicum* spp. against the pulse beetle, *C. maculatus*. Both extracts increased instant mortality and reduced F1 progeny. Repellency effects were 17–37% and 16–32% for petroleum ether and acetone extracts respectively.

vi. *Curcuma domestica* Linn. (Syn. *C. longa*) (Turmeric)

Family	Zingiberaceae
Distribution	India and South-East Asia
Part used	Rhizome
Chemical ingredients	Alkaloid, curcumin, essential oil, zingiberine.
Importance	Turmeric is a perennial plant with a short stem, and short and tufted leaves. It originated in

India and South-East Asia where there are deciduous monsoon forests. Now, it has worldwide distribution. It is an important spice and has considerable commercial importance as one of the principal ingredients of curry. The rhizome is the plant part being used in crop protection. It acts as an insecticide and repellent (Graigne, *et al.* 1985). The rhizome of the plant acts as good stored grain protectant against pests like rice weevil, rice flour beetle, grain borer, cowpea beetle, etc. (Jacobson, 1975; Pranata, 1985). In Srilanka, the turmeric roots are shredded with cow urine and water, and the solution used against caterpillars (Peries, 1986).

FIGURE 4.16   *CURCUMA DOMESTICA* LINN. (*SEE* PLATE 4.2)

vii. *Lantana camara* Linn. (Ghaneri/Gangutai)

Family	Verbenaceae
Distribution	India, widely distributed throughout the tropics and subtropics.
Part used	Leaves
Chemical ingredients	Essential oil, carvonene
Importance	Of the many known varieties, three have been reported from India of which *Lantana camara* Linn. (var. *aculeata* Moldenke) is most common. The occurrence of *L. camara* proper in India is doubtful. The plant is a prolific breeder, now a serious weed, occasionally used as green manure.

FIGURE 4.17   *LANTANA CAMARA* L. (*SEE* PLATE 4.4)

Leaves yield essential oil with a pleasant lasting odour reminiscent of sage (*Salvia officinalis* Linn), but it does not have any direct value in perfumery. Besides its insecticidal role, the plant is credited with vulnerary, diaphoresic, carminative and antispasmodic properties, used in fistulae, pustules, and tumours. Decoction given in tetanus, rheumatism and malaria and for ataxy of abdominal viscera. Some authors report its insecticidal nature.

1. Dua *et al.* (1996) evaluated the repellent effect of flowers against *Aedes* mosquitoes in the laboratory and field tests conducted with the plant area of BHEL (Bharat Heavy Electricals Limited) in Hadwar, UP (India). The flower extract in coconut oil provided 94.5% protection from *Aedes albopictus* and *Aedes aegypti*, single application can provide 50% protection for up to 4 h against possible bites of *Aedes* mosquitoes. They further reported that no adverse effects on the human volunteers were observed through the 3 months after application.

2. Meshram (2000) tested crude extract of the fresh leaves of different plant species including *Lantana camara* against 3rd instar larvae of the *Dalbergia sissoo* defoliator and *Placoptera reflexa* in order to evaluate antifeedant and insecticidal activity. The extract did have some antifeedant and insecticidal activity.

3. Misra (2000) treated black gram seeds with 3% dried leaf powder for control of *Callosobruchus chinensis* during storage in jars. The treatment was compared over five months with the local treatments of

red soil powder, cow dung, ash powder, etc. The seeds treated with plant powder completely prevented losses in seed weight and seed quality.

4. Deka *et al.* (2001) tested the efficacy of solvent extract of this plant against tea mosquito bug (*Helopeltis theivora*) in the field. The aqueous extract reduced the percentage of infestation from 27.63 to 38.90. However, chloroform extract gave the highest significant reduction (36.74 to 46.93%) followed by petroleum ether extract. All the concentrations of extract gave significant reduction of infestation but higher concentration of extracts gave better results.

viii. *Melia azedarach* Linn. (Bakain/Dharek/Chinaberry)

Family	Meliaceae
Distribution	Indian Himalayas, but today is widely distributed throughout the tropics and subtropics.
Part used	Seeds
Chemical ingredients	Tetranortriterpenoids (limnoids) related to azadirachtin
Importance	*Melia azedarach* Linn. commonly known as Dharek, Chinaberry tree, China tree, Persian lilac or 'Pride of India' is a close relative of neem tree. It acts as a contact and stomach poison. It shows insecticidal, repellent,

antifeedant and growth-inhibiting activity and is effective against ticks. Very strong antifeedant effect of the tree against locusts (*Schistocera gregaria*) was discovered during the locust invasion of Palestine in 1915. It was observed by an adolescent, Rachel Shpan–Gabrielith, that while all other vegetation was nearly completely devoured, the Persian lilac, was almost undamaged. Subsequently, laboratory studies established that the foliage remained untouched by locusts even after one week of starvation. Hot water extracts of the plant applied to wheat bran also prevented feeding by the locusts. Scientists subsequently used this technique successfully to save several crops during later invasions (1945, 1951) by locusts.

1. Islam *et al.* (1987) carried out laboratory and field studies to determine the optimal dose and application methods of this plant and to evaluate their antifeedant and growth-regulating properties. Methyl-tertiary butyl-ether (MTB) extract of this plant were most effective. The addition of sesame or linseed oil to MTB and methanolic extracts resulted in higher mortality of *Callosobruchus chinensis* and the Chrysomelid *Dicladispa armigera* and increased deterrent effect.

2. Schmutterer and Ascher (1987) observed a wide range of behavioural, physiological and toxic effects against several insect species.

3. Dhaliwal *et al.* (1996) tested the extracts in different solvents. These extracts as well as the pure compounds were found to exhibit phagodeterrent, oviposition-deterrent, fecundity- and longevity-reducing, development-disrupting and toxic effects.

4. Yanar *et al.* (2001) studied *in vitro*, the antifungal effect of the aqueous extract of this plant on pathogenic fungi like *Rhizoctonia solani* and *Phytophthora capsici*. The extract inhibits the growth of fungi.

FIGURE 4.18   *MELIA AZEDARACH* L. (*SEE* PLATE 4.5)

5. Castiglioni *et al.* (2002) reported the toxic effect of the aqueous extract of leaves, branches and fruits on the mite, *Tetranychus urticae*. It was also observed that they reduce the female mite fecundity.

6. Gajmer *et al.* (2002) studied the effect of methanolic extract of Neem (*Azadirachta indica* A. Juss) and bakain (*Melia azedarach* L.) seeds on oviposition behaviour and hatchability of eggs of *Earias vittella* (Fab.) There was no egg-laying when the moths fed on a sucrose diet containing 6, 8 and 10% neem and 10% bakain extracts.

7. Haryadi *et al.* (2003) tested the insecticidal properties of leaf powder against maize weevil, *Sitophilus zeamais*. The experiment was conducted under laboratory condition using diet in the form of artificial grains. To the diet 10 levels (0.0 to 4.5 % w/w) of the dry powdered plant material were added. 10 adults of unsexed weevil aged 7–15 days were added to 120 artificial grains in a closed but aerated plastic container. Each treatment was replicated 5 times, after 7 days of oviposition the parent insects were removed and discarded. The grains were incubated under laboratory condition and and after 3 weeks were examined for the emergence of progeny. The results showed that the presence of the plant materials in the diet significantly reduced the number of progeny, prolonged the developmental period and reduced the developmental index. Plant leaves in particular, showed strongly insecticidal effects. At 2% of the

leaf powder in the diet, the number of progeny in all 5 replications was reduced to zero.

8. Kokate and Chintalwar (2003) screened the petroleum ether extracts of this plant screened against pulse beetle, *Callosobruchus chinensis,* for repellent activity using an olfactometer. The extract showed repellency action.

ix. *Nicotiana tabacum* Linn. (Tobacco)

Family	Solanaceae
Distribution	South America (native), now cultivated all over the world
Part used	Leaves and stalk
Chemical ingredients	Nicotine and nor-nicotine alkaloids, anabasine, small amount of oxalic acid, glycoside, etc.
Importance	Tobacco plant is tropical in origin and thrives best in warm climate. It is grown under a wide range of conditions in tropical, subtropical and temperate zones. Specific varieties of *Nicotiana tabacum* are widely cultivated and used for cigarettes, bidi, cigar, cheroot, hookah, chewing and snuff.

Insecticidal use of nicotine dates back to the 17th century, long before chemists isolated and characterized nicotine as the major toxic principle of tobacco. During the 18th century tobacco in crude form as an aqueous extract or dust was employed as an insecticide in vegetable gardens in Europe. Commercial preparation of nicotine sulphate was sold in the market by 1910 and has been a popular insecticide ever since, because of its triple action as a stomach, contact and fumigant poison. Some earlier workers reported the insecticidal nature of this plant.

FIGURE 4.19   *NICOTIANA TABACUM* L. (*SEE* PLATE 4.6)

1. Horowitz, (1942) showed that efficiency of nicotine sulphate in controlling a variety of pests especially soft-bodied insects.

2. Isely (1944) stated that nicotine sulphate is safer, easier to handle and much less toxic to warm-blooded animals as compared to synthetic insecticides because of the high volatile nature of

nicotine. Insecticidal nicotine preparation leaves no appreciable residue on treated plants.

3. Carman *et al.* (1952) and Norton and Billings (1941) showed that nicotine (alkaloid) from this plant is believed to disappear in relatively short time usually in minutes thus leaving no hazardous material on marketable products. The above properties make the nicotine preparation a very ideal insecticide against the background of growing public concern over pesticide pollution.

4. Frear (1955) reported nicotine sulphate to be more toxic to insects than as only sulphate or hydrochloride.

5. Atwal (1976) recommended nicotine sulphate for combating apple wooly aphids. One of the advantages of the insecticidal use of nicotine is its reported high margin of safety for plants, causing little or no damage to the foliage.

6. Areekul *et al.* (1987) tested leaf extracts for their toxicity on 2-day-old adults of tephritid, *Dacus dorsalis,* in the laboratory at 25°C. The extract was moderately toxic (61–85%) causing mortality within 24 h.

7. Singh and Singh (1993) conducted laboratory experiments to investigate the efficacy of nicotine extract in the control of eggs and adults of *Chauliops fallax* on

soyabean. The result showed significant mortality of eggs as compared to control.

8. Archana Tiwary *et al.* (1995) isolated the alkaloid fractions from leaves and showed that they exhibited antifeedant activity against *Tribolium castaneum* adults and insecticidal and hormonal effects against larvae of *Tribolium castaneum*.

9. Amadioha and Obilor (2002) reported that it also acts as fungicidal plant. *Rhizopus oryzae* is an important pathogen of yam (*Dioscorea alata*), which causes soft rot diseases in stored yam tubers. The treatment with alcoholic extracts of leaves significantly reduces the radial growth of the pathogen *in vitro* and the spread of the rot disease *in vivo*. The extracts were more effective as a protectant than as curative biopesticides.

10. Madanlar *et al.* (2002) tested the pesticidal effects of this plant in the laboratory on various vegetable pests, and it was found to be promising.

x. *Pongamia glabra* L. (Syn. *Pongamia pinnata*) (Karanj)

Family	Leguminoceae
Distribution	India
Part used	Leaves and seeds
Chemical ingredients	Essential, bitter, fatty oil—karanjin, pongamol and glabrin

Importance

Pongamia glabra (Syn. *Pongamia pinnata*) variously known as karanja, puna oil tree, Indian beech or pongram.

The active components have been identified as karanjin, a furaflavone. Karanjin oil is applied as a surface protectant for effectively checking the infestation of pulse beetles and other stored grain pests. A concentration of 1% afforded complete protection even after 150 days and did not alter taste and smell of the grains. Pongamia cake was effective in controlling the attack of ground beetles on tobacco and did not leave any harmful residues in the soil. Some earlier reports include the following.

FIGURE 4.20　*PONGAMIA GLABRA* L. (*SEE* PLATE 4.7)

1. Ketkar and Schmutterer (1987) tested the non-edible oil from this plant against the bruchid stored product pest *Callosobruchus maculatus* on cowpea and *Callosobruchus*

chinensis on green gram. 1% oil gave complete protection to cowpea even after 150 days. 0.5% concentration of oil gave protection up to 90 days against *Callosobruchus chinensis* on green gram.

2. Chari and Ramaprasad (1993) revealed that the oil of Karanja repelled plant hoppers in rice and significantly reduced its ingestion and assimilation of food. Both brown plant hopper and white-backed plant hopper suffered heavy mortality but green leaf hopper was less susceptible.

3. Arora and Dhaliwal (1994) observed various types of bioactivity including its antifeedant, fecundity-curtailing and toxic properties.

4. Hiremath and Ahn (1997) obtained methanol extract by rotary vacuum evaporator at 40°C and tested against a pest of rice, the paddy brown plant hopper (*Nilaparvata lugens*). 39.1% mortality of adult female hopper was achieved.

5. Lohra *et al.* (2000) studied the effectiveness of biopesticides on the development of *Tribolium castaneum* Herbst. infesting stored sorghum. It was found that the treatment of grain with leaf extract of *Pongamia glabra* prolonged the development period of the insect pest from 5 to 15 days. A reduction in loss of

weight of sorghum treated with plant extract was also reported.

6. Meshram (2000) tested crude extract of fresh leaves against 3rd instar larvae of the *Dalbergia sissoo* defoliator, *Placoptera reflexa*, in order to evaluate their antifeedant and insecticidal activities. The investigations revealed that extract of this plant was the most effective and potent antifeedant and insecticidal agent.

7. Saminathan and Jayaraj (2001) evaluated the plant as a botanical pesticide against the cotton pest the mealy bug, *Ferrisia virgata* Cockerell. There was an increase in the percent mortality as time and dose increases.

xi. *Vitex nigundo* Linn. (Nirgundi/Nirgudi)

Family	Verbenaceae
Distribution	China, India
Part used	Leaves
Chemical ingredients	Alkaloids—nishidine and hydrocotylene
Importance	Nirgudi is a deciduous aromatic shrub 3 m tall. Flowers are lavender blue. There is a variety "incise" in this species, which is with deeply toothed leaves. It is native to China and India. It is commonly found in Ishyalee of Uttar Kashi between 500–1500 m. This species is also seen throughout the greater part of India and often grown for reclamation of forest land. Leaves are insecticidal and antibacterial.

Besides the insecticidal and antibacterial properties of leaves, it also has various other uses e.g., inhibiting action against *Mycobacterium tuberculosis*, anticancerous, tonic and vermifuge, smoked for relief in catarrh and headache, etc. The branches and leaves of the plant are insect-repellent and so are used for preservation of stored grains against insect attacks. A number of researchers reported the insecticidal properties of this plant, especially as an excellent stored grain protectant.

1. David *et al.* (1988) studied its pesticidal activity and chemical constituents. It has been shown by them to have repellent, insecticidal and juvenile hormone activity against several species of insect pests, including stored product pests, larvae of Lepidoptera and *Culex quinquefasciatus*.

FIGURE 4.21   *VITEX NIGUNDO* LINN. (*SEE* PLATE 4.8)

2. Khanam *et al.* (1990) tested methane extract of leaves against larvae of *Tribolium confusum* at various concentrations. At higher dose, greatest mortality, i.e., 93.81% after 20 days of application was obtained.

3. Sahayaraj and Sekar (1996) applied leaf extract to groundnut leaves to assess their efficacy against larvae of *Spodoptera litura*. Larval mortality was recorded after 96 h and it was 83%.

4. Meshram (2000) tested the crude extract of fresh leaves of this plant against 3rd instar larvae of the *Dalbergia sissoo* defoliator and *Plecoptera reflexa* in order to evaluate antifeedant and insecticidal activity. The investigation revealed that the extract had some antifeedant and insecticidal activity.

5. Misra (2000) studied the effectiveness of 3% dried leaf powder against the pulse beetle *Callosobruchus chinensis* L. on stored black gram. The results showed that there was reduced pulse beetle fecundity, and there was complete prevention of loss in seed weight and quality than the untreated control.

6. Anuradha *et al.* (2002) tested the insecticidal properties of this plant against *Callosobruchus maculatus* F. on cowpea (*Vigna unguiculata*), chickpea (*Cicer arietinum*) and green gram (*Vigna radiata*).

The observations showed more than 87% mortality of stored grain pest within 7 days.

7. Kalita *et al.* (2002) used this plant to control insect pests of stored pulses (grain legumes) against *Callosobruchus chinensis*. The result indicated that the plant was most effective in controlling pulse bruchid.

8. Mo-Jianchu *et al.* (2002) tested the toxic effect of leaf extract against larvae of *Culex pipiens* Pallens to provide a basis for the development of new chemicals for mosquito control. The ethanol extract showed different degree of toxicity at different concentrations. It was also observed that surviving larvae from the treatment group had longer developmental period and average pupa weight was higher than control group.

9. Sundararajan (2002) screened methanolic extract for their insecticidal activity against the 4th instar larvae of *Helicoverpa armigera* by applying dipping method of the leaf extracts at various concentrations on young tomato leaves. A larval mortality of more than 50% has been recorded for 2% test concentration within 48 h as compared with other plants. *Vitex nigundo* is found to have higher rate of mortality (82.5%) at 2% concentration.

10. Virendra Singh *et al.* (2002) screened isolated essential oil from leaves for insecticidal activity against *Sitotroga cerealella* infesting wheat seeds. The oil was effective against the pest and caused 100% mortality even at minimum concentration of 0.062%. Emergence of new adults was completely prevented by 0.125% dose concentration of essential oil.

11. Kokate and Chintalwar (2003) showed repellent activity of petroleum ether extract of this plant against pulse beetle, *Callosobruchus chinensis* Linn.

12. Patil and Goud (2003) evaluated the plant extract for its ovipositional repellent property against *Plutella xylostella* under laboratory conditions. The extract reduced the egg laying conditions at 24 h.

13. Tewary *et al.* (2003) evaluated the grain protectant potential of different bioactive parts of this plant against the stored product insect pest *Sitophilus oryzae*. All these bioactive parts were effective and act as good grain protectants reducing infestation.

REFERENCES

Alabi, O. and Olorunju, P.E. (2004). Evaluation of neem seed extract, black soap and cow dung for the control of groundnut leaf spot at Samaru, Nigeria. *Archives of Phytopathology and Plant Protection*. 37(2): 123–127.

Alavez-Solano, D., Reyes-Chilpa, J. R., Jimenez-Estrada, M.A., Villavicencio, M.A., Perez-Escandon, B.E. and Gomez-Garibay, F. (1998). Study on the insecticidal activity of seeds of *Pachirhizus erosus* (L.) urban. Proceeding of the 2nd International Symposium on Tuberosus Legumes Celaya; Guanajuato, Mexico. Augus 5–8, 535–548.

Al-Lawati, H.T., Azam, K.M. and Deadman, M.L. (2002). Insecticidal and repellent properties of subtropical plant extracts against pulse beetle, *Callosobruchus chinensis*. *Sultan Qaboos University Journal for Scientific Research*. 7(1): 37–45.

Allen, T.G., Dicke, R.G. and Harris, H.H. (1944). *Sabadilla, Schoenocaulon* spp. With reference to its toxicity to housefly. *J. Econ. Entomo*.37(3); 400–407

Amadioha, A.C. and Obilor, P.N. (2002). Fungitoxic effects of *Nicotiana tabacum* and *Cariya papaya* on *Rhizopus oryzae* causing yam tuber rot in storage. *Journal of Sustainable Agriculture and the Environment*. 4(1): 108–113.

Anonymous. (1928). Perris als Insektizid. *Tropenpflanzer*. 280–281.

Anonymous. (1977). Peasants against 'modern' technology. *Asian Action*. 11:4 (Thiland).

Anuradha, V., Gowari, N., Daniel, T. and Neelima, G. (2002). Effect of certain plant extracts on *Callosobruchus maculatus* as pest on stored pulses. *J. Ecobiol*. 14(1): 67–71.

Archana Tiwary, Kumar, M.L. and Saxena, R.C. (1995). Effect of *Nicotiana tabacum* on *Tribolium castaneum*. *Intern. J. Pharmacognosy*. 33(4): 348–350.

Areekul, S., Sinchaisri, P. and Tigavatan, A.S. Effect of Thai plant extracts on the Oriental fruit fly. *Kasetsart J. Nat. Sci*. 1987. 21(4): 395–407.

Arora, R. and Dhaliwal, G.S. (1994). Botanical pesticides in insect pest management. In: Dhaliwal, G.S. and Kansal, B.D., (eds.) *Management of Agricultural Pollution in India*. Commonwealth Publishers, New Delhi, India, 213–245.

Arthur, F.H. (1996). Grain protectant: Current status and prospects for the future. *J. Stored Prod. Res.* 32: 293–302.

Atwal, A.S. and Panji, H.R. (1964). Preliminary studies on the insecticidal properties of drupes of *Melia azedarach* against caterpillars of *Pieris brussicae. Indian J. Ento.* 26: 221–227.

Atwal, A.S. (1976). *Agricultural Pests of India and South East Asia*. Kalyani Publisher, New Delhi.

Attri, B.S. (1982). Neem as a source of pest control material. In: *Cultivation and Utilization of Medicinal Plants*. Atwal, C.K. and Kapur, B.M. (ed.)., RPL, CSIR, Jammu-Tawi.

Asolkar, L.V., Kakkar, K.K. and Chakre, O. J. (2000). *Glossary of Indian Medicinal Plants with active principles* (1965–81). Part- I (A-K) CSIR.

Barnes, D.K. and Freyre, R.H. (1969). Seed production potential of *Tephrosia vogelli* in Puerto Rico. Puerto Rico University. *Jour. Agr.* 53(3): 207–212.

Beroza, M.J. (1953). Killing agents. *Amer. Chem. Soci.* 75, 44 and 75, 2136.

Bestmann. H.J., Classen, B., Kobold, U., Vostrowsky, Klingauf, F. (1987). Botanical insecticides, IV. The insecticidal effect of the essential oil from the costmary Chrysanthemum, *Chrysanthemum balsmita* L. *Anzeiger-Fur Schadlingskunde, Pflanzenschutz, Umweltschutz.* 60:2, 31–34.

Bhatnagar Thomas, P.L. and Pal, A.K. (1974). Studies on the insecticidal activity of garlic oil. Differen toxicity of the oil to *Musca domestica* nebulo Fabr and Trog ma granarium Everts. *J. Food Science and Technology.* 11: 110–113.

Blommaert, K.L.J. (1950). The plant *Tephrosia vogelii* Hooker, as a fresh water fish poisons. *Roy. Soc. So. Africa Trans.* 32: 247–263.

Bloszyk, E., Szafranski, F., Drozdz, B. and Al-Shameri, K. (1995) African plants as antifeedant against stored product insect pests. *Journal of Herbs-Spices and Medicinal Plants.* 3:1, 25–36.

Boorsma, W.G. (1910). Pesticidal Plants. *Teysmannia.* 21: 624.

Bowers, W.S., Agregullin, M. and Garcia, E.S. (1987). Discovery and Identification of an antijuvenile hormone from *Chrysanthemum coronarium*. Proceedings of the International Symposium on insect Physiology, Biochemistry and Control. Rio de Janeiro, Brazil. Memorias-do-Instituto Oswaldo-Cruz. 82. Suppl. III, 51–54.

Busbey, R.L. (1939). A bibliography of quassia. U.S. Dept. Agr. Bur. Entomol and Plant Quarantine. E-483, 56.

Carman, G.E., Guenther, F.A., Blinn, R.C. and Garmus, R.D. (1952) *J. Econ. Entomol.* 45: 767.

Castiglioni, E., Vendramim, J.D. and Tamai, M.A. (2002). *Agrociencia-Montevideo.* 6(2): 75–82.

Chari, M.S. and Ramaprasad, G. (1993). Neem as an ancient and ovipositional repellent for *Spodoptera litura* F. In: Souvenir World Neem Conf., Feb. 24–28. Banglore, India. 81–90.

Cheng, T.H.(1945). *J. Econ. Entomol.* 38: 491.

Chitra, K.C., Reddy, P.S., Rao, P.K. and Goel, S.C. (1992). Efficacy of petroleum ether extracts of certain plants in the control of brinjal spotted leaf beetle, *Henosepilachna vigitioctopunctata* (Fabr.). Bioecology and control of insect pest symposium. 175–178.

Choudhary, R.K., Veda, O.P. and Mandloi, K.C. (2001). Use of Neem, *Azadirachta indica* (L.) and Garlic, *Allium sativum* (L.) in management of bollworm in cotton. Proceedings of the 88 session of the Indian Sci. Congress, Agricultural Sciences, New Delhi. 40–42.

Conucher, J. (1980). Pest, Predators and Pesticides, Organic Growers Association, W.A./ Australien.

David, B.V., Sukumaran, D. and Kandaasamy, C. (1988). The Indian privet *Vitex nigundo* Linn a plant possessing promising pesticidal activity. *Pesticides.* 22(2): 27–30.

Dayal, R., Tripathi, R.A. and Renu. (2003). Comparative efficacy of some botanicals as protectant against *Sitophilus oryzae* in rice and its palatability. *Annals of Plant Protection Sciences.* 11(1): 160–162.

Deb-Kirtanya, S. *et al.* (1980). Note on insecticidal properties of fruits of chilli. *Indian J. Agric. Sci.* 50(6): 510–512.

Deka, M.K., Karan Singh, Handique, R., Singh, K., (2001). Efficacy of wild sage (*Lantana camera* L.) and (*Adhatoda vasica*) against tea mosquito bug in the field. *Research on crops.* 2(1): 66–70.

Devraj, K. and Nandihalli, B.S. (2002). Efficacy of neem seed kernel dusts on pupae of *Helicoverpa armigera* (Hebner). *Insect-Environment.* 8(3): 107–108.

Dhaliwal, G.S., Arora, R. and Dilawari, U.K. (1996). Botanical pesticides in insect pest management: Emerging trends and strategies. In: Narwal, S.S. and Tauro, P. (eds). *Allelopathy in Pest Management for Sustainable Agriculture*. Scientific Publishers, Jodhpur, India. 93–119.

Downum, A.M. (1993). Using natural pesticides: Current and future perspectives. A report for the plant protection Improvement programme in Botswana, Zambia and Tanzania By Amelie Berger.

Dua, V.K. Gupta, N.C., Pandey, A.C, and Sharma, V.P. (1996). Repellency of *Lantana camera* (Verbenaceae) flowers against *Aedes* mosquitoes. *J. American Mosq. Control*. 12(3): 406–408.

El-Lakwah, F.A., Darwish, A.A. and Halawa, Z.A. (1996). Toxic effect of extracts and powders of some plants against the cowpea beetle (*Callosobruchus maculatus* F.) *Annals of Agricultural Science, Moshtohor*. 34(4): 1849–1859.

Frear, D.E.H. (1955). Insecticides derived from plants. In *Chemistry of Pesticides* D. van Norstrand Company, Inc., New York. 114–41.

Gajmer, T., Singh, R., Saini, R.K. and Kalidhar, S.B. (2002). Effect of methanolic extracts of neem (*Azadirachta indica* A. Juss) and bakain (*Melia azedarach* L.) seeds on oviposition and egg hatching of *Earias vittella* (Fab.) (Lep. Noctuidae). *J. Appl. Entomol*. 126(5): 238–243.

Gaskins, M.H., White, G.A. and Martin, F.W. (1972). *Tephrosia vogelii*: A source of rotenoids for insecticidal and piscicidal use. *U.S. Dept. Agr. Tech. Bull*. 1445: 38.

Georghiou, G.P. (1986). Pesticide resistance in time and space. In: Georghiou, G.P. and Saito, T. (eds.) *Pest Resistance to Pesticides*. Plenum Press, New York, USA. pp. 769–792.

Ghatak, S.S. and Bhusan, T.K. (1995). Evaluation on the Ovicidal activity of some indigenous plant extracts on rice moth, *Corcyra cephalonica* Stainton. (Galleridae: Lepidotera). *Environment and Ecology*. 13(2): 284–286.

Graigne, M., Ahmed, S., Mitchell, W.C. and Hylin, J.W. (1985). Plant species reportedly possessing pest control properties—An EWC/UH database. Resources System Institute, EWC Honolulu, College of Tropical Agriculture and Human Resources Uni. of Hawaii.

Greenstock, D. (1970). Garlic as a pesticide. HDRA Braintree, England.

Greenspan Gallo, L., Allee, L.L. and Gibson, D.M. (1996). Insecticidal effectiveness of *Mammae americana* (Guttiferaceae) extracts on larvae of *Diabrotica virgifera virgifera* (Coeleoptera:Chrysomelidae) and *Trichoplusta ni* (Lepidoptera: Noctuidae). *Economic Botany*. 50(2): 236–242.

GTZ. (1980). Post harvest problems Documentation of a OAU/GTS Seminar, Lome.

Gulati, B.C., Qureshi, N.A. and Taj-ud-din. (1982). Drug Research Laboratory, Jammu. Pyrethrum-Retrospect and Prospects. In: Cultivation and Utilization of Medicinal Plants edited by C.K. Atwal and B.M. Kapur, RRL, CSIR, Jammu-Tawi.

Gupta, R.K., Neerja Agrawal., Pandey, S.D., Srivastava, J.P. and Agrawal, R. (2001). Management of Mustard aphid *Lipaphis erysimi* Kalt using botanical pesticides. *Flora and Fauna, Jhansi*. 7(2): 107–109.

Gupta, M.P. (2005). Efficacy of Neem in combination with cow urine against mustard aphids and its effect on coccinellid predators. *Natural Product Radience*. 4(2).

Gu-YangFang, Shang-FuDe, Xue-JingZhong. (1997). The repellencies of six plant substances to adult *Tribolium castaneum* (Herbst). *Acta-agriculture-Universitatis-Henanensis*. 31(3): 277–279.

Hansberry, R. and Lee, C. (1943). *J. Econ. Entomol*. 36: 351.

Harish Chander, Ahuja, D.K., Nagender, A., Berry, S.K. and Chander, H. (1998). Efficacy of plant materials against *Tribolium castaneum* (Herbst) in milled rice under bagged conditions. *J. Insect Sci*. 11(2): 133–136.

Harpar, S.H., Potter, C. and Gillham, E.M. (1947). *Ann. Appl. Biol*. 34: 104.

Haryadi, Y., Yuniarti, S. and Credland, P.F. (2003). Study on the insecticidal effects of custard apple (*Annona reticulate*) and Mind (*Melia azedarach*) leaves against *Sitophilus zeamais* M. Adv. in stored product protection, York UK. 600–602

Heal, R.E., Rogers, E.F., Wallace, R.T. and Starnes, O. (1950). A survey of plants for insecticidal activity. *Lloydia*. 13: 89.

Hiremath, T.G. and Ahn, Y.J. (1997). Parthenium as a source of pesticide. 1st International Conference on Parthenium Management. Univ. Agric. Sci. Dharwad. pp. 86–89.

Horowitz, B. (1942). Australian *Journal of Sci*. 4: 179.

Ho, S.H., Koh, L., Ma, Y., Huang, Y., Sim, K.Y. (1997). The oil of Garlic, *Allium sativum* L. (Amaryllidaceae), as a potential grain protectant against *Tribolium castaneum* (Herbst) and *Sitophilus zeamais* Motscb. Post harvest Biology and Technology. 9(1): 41–48.

Isely, D. (1944). *Methods of Insect Pest Control*, Part II, 3rd edn. Burgess Pub. Co., Minneapolis, Minnesota.

Islam, B.N., Schmutterer, H. and Ascher, K.R.S. (1987). Use of some extracts from Meliaceae and Annonaceae for control of rice hipsa *Dicladispa armigera* and the pulse beetle, *Callosobruchus chinensis*. Proceedings of third Neem Conference, Nairobi, Kenya. 217–242.

Iwuala, M.O.E. (1981). Denetta oil, A potential new insecticides: Tests in the adults and nymphs of *Periplaneta americana* and *Zonocerus varriegatus*. *J. Econ. Entomol.* 74(3): 249–252.

Jacobson M. and Crosby, D.G. (1971). *Naturally occurring insecticides*. Marcel Dekker, Inc., New York.

Jacobson, M. (1975). Insecticides from plants. A review of the literature. 1954–1971. *Agriculture Handbook*. No. 461. USDA. Washington DC.

Jacobson, M. (1990). *Glossary of Plant Derived Insect Deterrent*. CRC Press, Boca Raton, Florida, USA.

Jeyabalan, D. and Murugan, K. (1997). Effect of neem limnoids on feeding and reproduction of *Helicoverpa armigera* (Hubner) (Lepidoptera: Noctuidae). *Entomon*. 22 (1): 15–20.

Jha, M.M., Sanjeev Kumar and Saleh Hasan. (2004). Effects of botanicals on Maydis leaf blight of Maize *in vitro*. *Annals of Biology*. 20(2): 173–176.

Jung, K. (1938). Pflanzliche Insektizide (Pyrethrum, Derris, Mundulea, Lonchocarpus and Tephrosia). *Tropenpflanzer*. 411: 431–443.

Kalita, J., Gogoi, P., Bhattachraya, P.R. and Handique, R. (2002). A study on effect of traditional storage protectants on oviposition of *Callosobruchus chinensis* and their damage to stored pulses. *J. Appl. Zool. Res.* 13(2-3): 251–254.

Karbasayya, Rahamn, M.F. (2001). Efficacy of certain neem products and pesticides as seed treatment for the management of *Meliodogyne incognita* on black gram.

Ketkar, C.M. and Schmutterer, R. (1987). Use of tree derived non-edible oils as surface protectants for stored legumes against *Callosobruchus maculatus and C. chinensis*. Nat. pesticides from neem tree and other tropical plants. Proceedings of the 3rd international Neem Conference, Nairobi, Kenya. 535–542.

Khanam, L.A.M., Talukder, D. and Khan, A.R. (1990). Insecticidal property of some indigenous plants against *Tribolium confusum* Duval (Coleoptera: Tenebrionidae). *Bangladesh J. Zool.* 18(2): 253–256.

Knapp, M., Kashenge, S.S. and Knapp, M. (2003). Effects of different neem formulations on the two spotted spider mite, *Tetranychus urticae* Koch, on Tomato (*Lycopersicon esculentum* Mill.). *Insect Science and its Application.* 23: 1, 1–7.

Kokate, S.D. and Chintalwar, G.J. (2003). Insect repellent activity of certain plant extracts against pulse beetle, *Callosobruchus chinensis* L. *Nat. Acad. Sci. Letters.* 26(1–2): 44–46.

Kumar, B.H. and Thakur, S.S. (1988). Certain non–edible seed oils as feeding deterrents against *Spodoptera litura* Fb. *Journal of the Oil Technologists Association of India.* 20(3): 63–65.

Latif, Z., Craven, L., Hartley, T.G., Kemp, B.R., Potter, J., Rice, M.J., Waigh, R.D., Waterman, P.G. (2000). An insecticidal quassinoid from the new Australian species *Quassia* sp. Aff. Bidwillii. Biochemical Systematics and Ecology. 28(2): 183–184.

Lee, C.S. and Hansberry, R. (1943). Toxicity studies of some chinese plants. *J. Econ. Entomol.* 36: 915.

Lohra, Y., Singhvi, P.M. and Lohra, Y. (2000). Effectiveness of biopesticides on the development of *Tribolium castaneum* Herbst. Infecting stored sorghum. *J. Appl. Zoo. Res.* 11 (2–3): 128–131.

Luo-Du Qiang, Zhang-Xing, Tian-Xuan, Liu-Jikai. (2004). Insecticidal compounds from *Tripterygium wilfordii* active against Mythiman separata. Zwitschrift-fur-Naturforschung-Section-C. *Biosciences.* 59(5/6): 421–426.

Madanlar, N., Yoldas, Z., Durmusoglu, E. and Gul, A. (2002). Investigations on the natural pesticides against pests in vegetable greenhouses in Izmir (Turkey). Turkiye-Entomoloji-Dergisi. 26(3): 181–195.

Maheshwari, V.L., Hemlata Kotkar, Mendki, P.S. and Shipra, R. Jha. (2001). Antimicrobial and pesticidal activity of partially purified flavonoids of *Annona squamosa. Pest Management Sci.* 58: 33–37.

Mallapur, C.P., Hulihalli, U.K. and Kubsad, V.S. (2001). Safflower aphid management through botanical insecticides. *Karnataka J. Agri. Sci.* 14(2): 321–325.

Mambelli, P., Marchini, B., Bazzocchi, C., Pari, P., Tellarini, S. (1996). Preliminary studies on optimizing the utilization of quassia wood (*Quassia amara* L.; *Picrasma excelsa* (Swz) Lindl.) as a biological insecticide. Atti Convegno internazionale: coltivazione e miglioramento di piante officinali, Trento, Italy. 2(3): 317–324.

Mancebo, F., Hilje, L., Mora, G.A., Salazar, R. (2000). Antifeedant activity of plant extracts on *Hypsipyla grandella* larvae. Revista-Forestal-Centroamericana. 31: 11–15.

Mathur, A.C. and Saxena, B.P. (1975). Introduction of sterility in Male Houseflies by vapours of *Acorus calamus* L. oil. *D. Naturwissenschaften*. 12: 576.

McKeen, C. (1956). The inhibitory acitivity of extract of *Capsicum frutescens* on plnat virus infections. *Canadian J. Botany*. 34: 891–903.

Meshram, P.B. (2000). Antifeedant and insecticidal activity of some medicinal plant extracts against *Dalbergia sissoo* defoliator *Plecoptera reflexa* Gue (Lepidoptera: Noctuidae). *Indian-Forester*. 126(9): 961–965.

Minja, E.M., Silim, S.N., Karuru, O.M. (2002). Efficacy of *Tephrosia vogelli* crude leaf extract on insects feeding on Pigeon pea in Kenya. *International Chick Pea and Pigeonpea Newsletter*. 9: 49–51.

Misra, H.P. (2000). Effectiveness of indigenous plant products against the pulse beetle, *Callosobruchus chinensis* on stored black gram. *Ind. J. Entomol*. 62(2): 218–220.

Mo-JianChu, Teng-Li and Yang-TianCi. (2002). Toxic effects of the ethanol extract of several garden plants against the third instars larvae of *Culex pipiens* Pallens. *Chinese Journal of Vector Biology and Control*. 13(5): 327–329.

Nandgopal, V. and Ghewande, M.P. (2004). Use of Neem products in groundnut pest management in India. *Natural Product Radience*. 3 (3).

Neem: A tree for Solving Global Problems. Report of Adhoc Panel of the board on Science and technology for International Development, National Research Council, National Academy Press, Washington, DC, USA. 1992.

Norton, L.B. and Billings, O.B. (1941). Characteristics of different types of nicotine sprays: I. Nicotine residues. *J. Econ. Entomol*. 34(5): 630.

Osario, L.G. (2004). Plant protecting other plants—an alternative to pest resistant GM crops. *MFP News.* 14(1): 3–4.

Park, C. Kim, S.I. and Ahn, Y.J. (2003). *Insecticidal activity of asorones identified in Acorus gramineus rhizome against three Coleopteran Stored Product Insects Res.* 39(3): 333–342.

Parmar, B.S. and Ketkar, C.M. (1993). In: *Neem Research and Development.* Randhawa, N.S. and Parmar B.S. (eds). Society of Pesticide Science, New Delhi, India. 270–283.

Patil, R.S. and Goud, K.B. (2003). Efficacy of methanolic plant extracts as ovipositional repellents against diamond back moth, *Plutella xylostella* L. *J. Entomol. Res.* 27(1): 13–18.

Perez, M.P., Pascaul Villalobus, M.J. (1999). Effect of the essential oil of flower buds of *Chrysanthemum Coronarium* L. on white fly and stored pests. *Investigacion-Agraria. Produccion y Proteccion Vegetale.* 14(1/2): 249–258.

Peries, L. (1986). *Cattle urine as a substitute for agrochemicals: Nat. Rural Confer.*

Plank, H.K. (1950). Federal Expt. Std. Puerto Rico Mayaguez. 49 (1).

Plummer, C.C. (1938). U.S. Dept. Agri. Circ. 455.

Pranata, R.I. (1985). Possibility of using turmeric (*Cucurma longa* L.) for controlling storage insects. *Bio. Newsletter.* Bogor/Indonesia.

Prasad, K.D., Ahmad, M.A., Ansari, M.Z. and Kumar, A. (1996). Effect of *Annona squamosa* seed on *Culex* larvae. *Ind. J. Indigenous Med.* 17(1): 83–87.

Punam Kumari, Prem Kumari, Verma, M.K. and Kumari, P. (1999). Effect of *Acorus calamus* as pulse grain protectant against *Callosobruchus chinensis. J Appl. Biol.* 9(2): 188–189.

Rajsekaran, T., Pereira, J., Ravishankar, G.A., Venkataraman, L.V. (1996). Repellency of callus derived pyrethrin to mosquito *Culex quinquefasciatus* Say and red flour beetle *Tribolium castaneum* Herbst. *International Pest Control.* 38(5): 154–156.

Rao, N.S., Raguraman, S. and Rajendran, R. (2003). Laboratory assessment of the potentiation of neem extract with the extracts of sweet flag and pungam on bhendi shoot and fruit borer, *Earias vitella* (Fab). *Entomon.* 28(3): 277–281.

Roark, R.C. (1947). Some promising insecticidal plants. *Econ. Botany.* 1: 437–445.

Rogers, E.F., Snyder, H.R. and Fischer, R.F. (1952). *J. Amer. Chem. Soc.* 74.

Sahayaraj, K. and Sekar, R. (1996). Efficacy of plant extracts against tobacco caterpillar larvae in groundnut. *International Arachis Newsletter*. 4: 38.

Saminathan, V.R. and Jayaraj, S. (2001). Evaluation of botanical pesticides against the mealybug, *Ferrisia virgata* Cockerell (Homoptera: Pseudococcidae) on cotton. *Madras Agriculture Journal*. 88(7–9): 535–537.

Schmutterer, H. and Ascher, K.R.S. (eds). (1987). Natural pesticides from Neem Tree and other Tropical plants. Proceedings 3rd International Neem Conference, July 10–15, 1986, Nairobi, Kenya.

Schmutterer, H. (1990). Properties and potential of natural pesticides from the neem tree *Azadirachta indica*. *A Rev. Ent.* 35: 271–297.

Schmutterer, H. and Singh, R.P. (1995). List of insect pests susceptible to neem products. In: H. Schmutterer (ed.). *The Neem tree. Source of unique Natural Products for integrated pest management, Medicine, Industry and other purposes. Azadirachta indica* A Juss and other Meliaceae plants. VCH; Weinheim Germany. pp. 326–365.

Shepard, H.H. (1951). *The Chemistry and Action of Insecticide*. Mc Graw-Hill, New York.

Singh, K.J. and Singh, O.P. (1993). Efficacy of biopesticides on egg and adults of cowpea pod bug (*Chaullops fallax*) infesting soybean (*Glycine max*). *Ind. J. Agri. Sci.* 63(11): 756–758.

Singh, Ram and Reena. (2003). Insecticidal properties of garlic, *Allium sativum*— A review. *J. Med. and Aromatic plant Sci.* 25(4): 1024–1038.

Singh, R.P. and Singh, S. (1996). Neem for Management of Insect Pests: Advantages and Disadvantages. Lal, O.P. (ed.). *Recent Advances in Indian Entomology*. APC Publications Pvt. Ltd., New Delhi. 67–82.

Singh, H., Mrig, K.K. and Mahla, J.C. (1996). Efficacy and persistence of plant products against lesser grain borer, *Rhyzopertha dominica* (F.) in wheat grain. *Annals of Biol.* (Ludhiana). 12(1): 99–103.

Spickett, R.G.W. (1955). The chemistry of some lesser known insecticides of plant origin. Colonial Plant and Animal products. 5: 288–303.

Subramanian, T.V. Sweet flag (*Acorus calamus*)—A potential source of valuable insecticide. *J. Bombay Nat. Hist. Soc.* 1948/49. 48: 338–341.

Sundaranjan, G. (2002). Control of caterpillar *Helicoverpa armigera* using botanicals. *Journal of Ecotoxicology and Environmental Monitoring*. 12(4): 305–308.

Ta-wang, Y. (1941). *Chinese Med. J.* 60: 222.

Tewary, D.K., Santosh and Vasudevan, P. (2003). Bioefficacy of *Vitex nigundo* L. (Verbanaceae) plant products as protectant against *Sitophilus oryzae* (L) (Coleoptera: Curculionoidae). *Shashpa*. 10(2): 161–166.

Tripathi, A.K., Veena, Prajapati, Sushil Kumar, Prajapati, V. and Kumar, S. (2003). Bioactivities of L-carvone, D-carvone and dihydrocarvone toward three stored productbeetles. *J. Econ. Entomol.* 96(5): 1594–1601.

Ujvary, I., Casida, J.F. (1997). Partial synthesis of 3-O Vanilloylveracevine, an insecticidal alkaloid from *Schoenocaulon officinale*. *Phytochemistry*. 44(7): 1257–1260.

Unjitwatana, U., Kochrat, S., Sangwanich, A., Udomluck-Unjitwatana, Somuegkochrat and Arom-Sangwanich. (2003). Acute toxicity of derris (*Derris elliptica* B.) and its effect on cholinesterase of *Tilapia nilotica* L. Proceedings of 41st Kasetsart University, Bangkok, Thailand. Annual Conference, Fisheries, 298–303.

Van Latum. Using natural pesticides: Current and future perspectives. A report for the plant protection Improvement programme in Botswana, Zambia and Tanzania Edited by Amelie Berger. 1991.

Verma, J. and Dubey, N. (1999). Prospective of botanical and microbial products as pesticides of tomorrow. *Curr. Sci.* 76: 172–179.

Vijayalakshmi Majumdar, Mishra, S.D. and Mojumdar, V. (2001). *Curr. Nematology*. 12(1-2): 7–10.

Virendra Singh; Rameshwar Dayal; Bhandari, R.S., Singh, V. and Dayal, R. (2002). Pesticidal activity of essential oil of *Vitex nigundo* leaves against *Sitotroga ceralleata*. *Shashpa*. 9 (1): 71–75.

Visetson, S., Milne, M., Suraphon Visetson, Manthana Milne. (2001). Effects of Root extract from Derris (*Derris elliptica* Benth) on mortality and detoxification enzyme level in the Diamondback moth larvae (*Plutella xylostella* Linn). *Kasetsart J. Nat. Sci.* 35(2): 157–163.

Vyas, B.N., Ganesa, S., Ramna, K., Godrej, N.B., Mistry, K.B., Singh, P.P. (1999). Effect of Three Plant Extracts and Achook: a Commercial Neem Formulation on

Growth and Development of Three Noctuid Pests. Science Publishers, Inc. Enfield USA. 103–109.

Yadava, R.L. (1971). Use of essential oil of *Acorus calamus* L. as an insecticide against the pulse beetle *Callosobruchus chinensis* L–Z. *Angew., Ent.* 68: 289–294.

Yadav, J.P. and Bhargava, M.C. (2002). Effect of certain botanical products on biology of *Corcyra cephalonica* Stainton. *Ind. J. Plant Protect.* 30(2): 207–209.

Yamasaki, R.B., Ritland, T.G., Barnby, M.A., Klocke, J.A. (1988). Isolation and purification of Salannin from neem seeds and its quantification in neem and chinaberry seeds and leaves. *J. Chromatography.* 447(1): 277–283.

Yanar, Y., Kadioglu, I., Kutluk, N.D., Cesmeli, I. and Hangun, A. (2001). *Turkiye Harboliji-Dergisi.* 4 (1): 58–63.

Zhang Ye Guang, Xu-Hanttong, Huang-JiGuang, Chiu-ShinFoon, Zhung-YG, Xu-HH, Huary-J.G. and Chiu, S.F. (2000). The antifeedant activity of *Tephrosia vegelli* (Hook) against species of Lepidoptera. *J. South China Agri. Uni.* 24(4): 26–29.

5

A SURVEY OF BIOCIDAL PROPERTIES OF SOME INDIGENOUS PLANTS

5.1 INTRODUCTION

A survey of 60 indigenous plants from 31 families used as biocidal plants and reported by earlier researchers was undertaken for review. Among the total families, some important families, contribute more number of plants, e.g., Asteraceae (6), Leguminosae (5), Solanaceae (5), Verbenaceae (5), Euphorbiaceae and Apocynaceae (3), etc. Detailed study on the known active ingredients indicates that saponin, alkaloid and glucosides are found in more plants, followed by essential oil, rotenone and other allied substances. These plant-based botanical products have been utilized by man since ancient times either for controlling the population of harmful pests like insects, or for catching fishes for consumption and in therapeutics as antimicrobial agents (Asolkar *et al.*, 2000; Dhaliwal and Arora, 2003). Here an attempt has been made to compile available data on 60 indigenous plants. These act as piscicidal, insecticidal, antibacterial and antifungal agents and

TABLE 5.1   LITERATURE SURVEYS OF SOME INDIGENOUS PLANTS AS A BIOCIDAL AGENTS

No.	Botanical and vernacular name with family	Chemical ingredients	Biocidal activity reported					Medicinal uses
			Piscicidal	*Insecticidal	Anti-bacterial	Anti-fungal	Others	
01	*Acacia concinna* DC Shikekai Leguminoceae	Saponin, alkaloid	–	91, 182	–	–	Anthelmintic, nematocidal	Pods–laxative, anti-diarrhoeal.
02	*Acorus calamus* Linn Vekhand Araceae	Glucoside, acorin, essential oils, alcohols	–	2, 19, 27, 31, 32, 36, 37, 38, 39, 50, 55, 72, 78, 79, 88, 90, 91, 124, 126, 139, 141, 146, 151, 158, 180, 182, 212, 236	180	19, 159	Anthelmintic	Nerve tonic, anticancerous hypothermic, hypotensive.
03	*Adhatoda vasica* Nees Adulsa Acanthaceae	Alkaloid–vascine, essential oil	–	15, 18, 37, 49, 90, 182, 184, 209	110, 179, 221	159	–	Antiseptic, anti-fertility, antiviral, anti-tubercular, wound healing, anti-inflammatory.
04	*Agave americana* Linn Kekati Amaryllidaceae	Saponins, agave-saponin, piscidic acid	36, 37, 112	36, 38, 57, 79, 130	–	–	–	Used in scurvy, syphilis and gonorrhoea, antiseptic, laxative.
05	*Allium sativum* Linn Lasoon/Garlic Liliaceae	Diallyl sulphide, diallyl trisulphide	19	19, 23, 40, 55, 87, 90, 105, 160, 184, 197, 205, 212	6, 19, 47, 63, 110, 179, 180, 219	19, 44, 59, 63, 89, 105, 131, 154, 166, 188, 201, 226, 238	Nematocidal, anthelmintic	Anti-tubercular, anti-diabetic, anti-cancerous, antiseptic, carminative, hypotensive, rheumatism.
06	*Anagallis arvensis* Linn Jonkmari Primulaceae	Saponin, glycosidic saponin, enzyme	19, 36, 37, 38, 91, 112	148	–	19	Anthelmintic, wormicidal	Toxic to sheep, used in dropsy, gout, rheumatism, snake bite, hepatic and renal complaints, antiviral.
07	*Anamirta cocculus* Linn Kakmari Menipermeaceae	Picrotoxin, cocculin, anamirtin	25, 37, 91	37, 83, 91	–	91	Anthelmintic	Ointment on chronic skin diseases. Thermogenic, used in bronchitis, foul ulcers and ringworm. Anticancerous.

No.	Botanical name / local name / family	Active principles					Biocidal property	Uses
08	*Andropogan schoenthus* Linn Rohisha gavat Poaceae	Essential oil, geraniol, citronellal	–	5, 19, 74, 101, 120, 152	19, 74, 174, 232	10, 19, 74, 174	Nematicidal	Antiseptic, antimicrobial, used in elephantiasis.
09	*Annona muricate* Linn Hanuman phal Annonaceae	Alkaloid, essential oil	32, 37, 112	32, 37, 74, 134	–	–	Anthelmintic, molluscicidal.	Astringent, parasiticide.
10	*Annona squamosa* Linn Sitaphal Annonaceae	Alkaloid, toxic resin and oil	19, 36, 37, 91, 111, 112	8, 17, 19, 32, 34, 36, 37, 38, 50, 55, 72, 74, 80, 82, 83, 85, 86, 91, 102, 114, 122, 126, 134, 146, 151, 180, 182, 184, 209, 225, 230	114	114, 204	–	Antiseptic, abortifacient, wound healing, anti-cancerous, used in mental depression and spinal disorders.
11	*Azadirachta indica* A. Juss Neem Meliaceae	Azadirachtin, bitter oil, margosic acid, nimbin and nimbidin	19, 135	3, 8, 15, 18, 19, 24, 25, 30, 36, 38, 40, 45, 50, 51, 54, 55, 57, 68, 72, 80, 82, 83, 90, 93, 95, 104, 105, 106, 116, 120, 122, 126, 129, 139, 141, 146, 151, 162, 164, 170, 173, 176, 180, 184, 193, 198, 199, 203, 209, 214, 217, 225, 230, 237, 239	19, 39, 53, 110, 179, 180, 213	10, 19, 33, 59, 89, 99, 105, 149, 166, 180, 190, 196, 204, 226	Wormicidal, nematicidal, acaricidal, toxic to frog tadpole.	Antiseptic, anti-diabetic, in skin disease, spermicidal, anti-cancerous, antiviral, used in snake bite and scorpion sting, anti-tubercular.
12	*Balanites roxburghii* Planch Hinganbate Simaroubaceae	Saponin, steroidal, sapogenin	19, 36, 37, 38, 12, 148, 150	203	19, 53	–	Anthelmintic	Used in snake bite and leucoderma.
13	*Barringtonia racemosa* Linn Samudraphal Lecithidaceae	Glucoside, saponin, barringtonin, tannin	2, 31, 36, 37, 38, 39, 83, 91, 112, 150, 182	31, 37, 39, 83, 91, 182, 234	97	–	–	In jaundice, seeds aromatic used in colic and ophthalmia.
14	*Brassica nigra* Linn Mohari/Rai Cruciferae	Glucoside, essential volatile oil	–	19, 24, 45, 83, 120, 141, 199, 207, 214, 225	–	–	–	Snake bite, rheumatism, constipation, antiseptic.

(*Contd.*)

TABLE 5.1   (*CONTINUED*)

No.	Botanical and vernacular name with family	Chemical ingredients	Biocidal activity reported					Medicinal uses
			Piscicidal	* Insecticidal	Anti-bacterial	Anti-fungal	Others	
15	*Butea monosperma* Lam Palas Leguminosae	Glucoside, butrin, butein, butin, etc.	–	18, 80	–	–	Anthelmintic, wormicidal	Seeds astringent in diarrhoea, dysentery and depurative, aphrodis. Used in snake bite, contraceptive and night-blindness.
16	*Capsicum annum* Linn Mirchi/Red Chilli Solanaceae	Capsicin, volatile alkaloid, solanine	–	62, 70, 87, 126	33, 47	–	–	Used in sore throat hoarseness, dyspepsia, yellow fever, diarrhoea, piles, snake bite, carminative, local irritant.
17	*Citrullus colocynthis* Shrad Kadu-Indravan Cucurbitaceae	Bitter substances, colocynthin, saponin, glucoside, tannin	–	15, 63, 156, 162, 193, 205	–	–	Anthemintic	Anti-tubercle, rheumatism, jaundice, snake bite, in Kala-azar, hypoglycemic, antiseptic, anti-poisoning.
18	*Clerodendrum phlomidis* Linn. F. Arni Verbenaceae	Bitter substances	–	62	–	189	Wormicidal	Used in measles, syphilitic complaints, aromatic, astringent.
19	*Clerodendrum serratum* Linn. M. Bharangi Verbenaceae	Bitter substances, alkaloid, saponin, glucoside	–	146	–	–	Anti-protozoal, anthelmintic	Used in snake bite, malaria, rheumatism, epilepsy, remedy for respiratory diseases.
20	*Cocculus hirsutus* (Linn) Diel Vasanvel Menispermeaceae	Alkaloid, coclaurine, magnoflorine, trilobine	–	37, 83, 231	132	132	–	Cooling medicine for gonorrhoea, eczema, impetigo, cardiotonic, anti-tubercular.
21	*Coriandrum sativum* Linn Kothmir/Dhaniya Apiaceae	Essential oil-coriandrol	–	65, 67	19	19	–	Tonic, anti-tubercular, in spleen complaints, venereal disease syphilis.

22	*Croton tiglium* Linn Jamalgota Euphorbiaceae	Toxic resin, oils, proteins—croton-I, lectin, β-sitosterol	19, 32, 36, 37, 83	17, 19, 38, 83, 184	–	227	–	Snake bite, carcinogenic, abortifacient, oil pergative, anticancerous.
23	*Curcuma longa* Linn Halad Zingiberaceae	Curcumin, alkaloid, essential oil	–	5, 19, 36, 37, 45, 74, 78, 87, 151, 157, 172	19, 71, 74, 75, 110, 220	19, 71, 75, 94, 154	Anti-amoebic, (anti-entamoeba). Nematicidal, anthelmintic.	Anti-parasitic, blood purifier, anti-fertile agent, antihepatotoxic, wound hea-ling, anti-inflammatory, anti-poisonous, anti-ulcer.
24	*Datura metal* Linn Dhotra Solanaceae	Alkaloid, atropine, hyoscyamine	150, 182	18, 60, 71, 79, 91, 102, 225, 239	–	42, 44, 99, 149, 208, 226	Nematicidal, germicidal	Insanity, anti-dandruff, seed—in hydrophobia, malaria, antiseptic, treating of Rabid dog.
25	*Dodonaea viscose* (Linn) Jac. Didoni Sapidaceae	Saponin, alkaloid, glucoside, dodonin	36, 37, 38, 39, 112	1	19	–	Anthelmintic	Used in swelling, burns, wound, rheumatism, hypotensive, spasmolytic, sedative.
26	*Duranta repens* Linn Duranda Verbenaceae	Saponin, alkaloid (Narcotine)	–	36, 37, 38, 62, 147	147	19, 147	–	Hedge plant.
27	*Echinops echinatus* Roxb Katechubak Asteraceae	Echinopsine	–	2, 32, 37, 91, 182	–	–	–	Germicidal, antidotes to scorpion sting.
28	*Eucalyptus globules* Labill Nilgiri Myrtaceae	Essential oil, cineole, sesquiterepenes, alcohol, volatile oil	–	19, 36, 37, 38, 51, 62, 101, 139, 152, 162, 193, 194, 239	19, 180	59, 99, 149, 194	–	Antiseptic, anti-malarial, used as expectorant in chronic bronchitis.
29	*Euphorbia tirucalli* Linn Sher/Thorya Euphorbiaceae	Latex contain euphorbone, tavaxerol, α-euphorol	36, 37, 38, 91, 112, 150	17, 235	107	19, 154	–	Warts, asthma, in snake and scorpion bite, rheumatism.
30	*Gardenia lucida* Roxb. Dikamali Rubiaceae	Essential oil, bitter substances, resin.	–	37, 42, 182, 212	–	–	Anthelmintic	Keep off flies and worms, antiseptic, antimicrobial.

(Contd.)

TABLE 5.1 (*CONTINUED*)

No.	Botanical and vernacular name with family	Chemical ingredients	Biocidal activity reported					Medicinal uses
			Piscicidal	* Insecticidal	Anti-bacterial	Anti-fungal	Others	
31	*Gloriosa superba* Linn Khadyanag/Buchnag Liliaceae	Alkaloid—superbrine, gloriosine, colchicines.	19, 91,150	36, 38, 182	–	–	–	Leprosy, parasitic affliction of skin, snake bite, scorpion sting, oxytocic effect on uterus.
32	*Jatropha curcas* Linn Mogali errand Euphorbiaceae	Toxic principle curcin, fatty oil and resinous matter.	36, 37, 38, 150, 176, 181	50, 68, 91, 191, 230	–	33	Anthelmintic	Juice used for scabies, eczema, ringworm, rheumatism, paralytic afflictions, antidiarrhoea.
33	*Lantana camara* Linn Ghaneri Verbenaceae	Essential oils—camerene, micranene and isocamerene.	–	37, 49, 60, 68, 74, 79, 96, 116, 122, 126, 139, 141, 145, 164, 183, 184, 191, 225, 228, 229	48, 78, 175, 229	48, 59, 202, 226, 229	Molluscicidal, nematicidal	In tetanus, malaria and rheumatism.
34	*Lycopersicon esculentum* Mill. Tomato Solanaceae	Oxalic acid, narcotine and solanine	–	146, 206	–	–	–	Keep away poison, used in asthma, bronchitis, and appetizer.
35	*Madhuca indica* J.F. Gmel Mahua Sapotaceae	Alkaloid, glucosidic saponin, β-sitosterol	18, 36, 37, 91, 111, 112, 182	15, 18, 36, 38, 68, 79, 80, 91, 101, 133, 162, 164, 176, 182, 199, 230	25	226	–	Fomentation, piles, tonic, nutritive appetizer, anti-tubercular, rheumatism.
36	*Melia azadirachta* Linn Bakam /Indian Illiac Meliaceae	Alkaloid, azadidine, resin, tannin	–	20, 55, 65, 80, 82, 83, 90, 100, 106, 122, 141, 146, 147, 152, 181, 193, 237	192, 213	33, 226, 238	Anthelmintic, nematicidal	Antiseptic, in rheumatism, antiviral, acaricidal.
37	*Mentha arvensis* Linn Pudina Lamiaceae	Essential oil—D-carvone, carene and citronellol	–	45, 65, 100, 101, 125, 181, 224	232	33, 44, 140, 189	–	Rheumatism, carminative, antiseptic.
38	*Murraya oenigii* Spreng. Kadhipatta Rutaceae	Essential oil, glucoside, koeinigin	–	4, 78, 139, 144	41, 213, 221	–	–	Anti-poisonous, anti-dysentry, anti-vomiting.

39	*Nerium oleander* Linn Kanher Apocynaceae	Glucoside, neriin and oleandrin	222	22, 25, 50, 55, 83, 181	1 86	186	–	Poisonous tree, cardiotoxic to heart of vertebrates.
40	*Nicotiana tabacum* Linn Tobacco Solanaceae	Glucoside, alkaloids-nicotine, nicoteine, nicotimine, anabasine	37, 112, 182	16, 17, 25, 36, 38, 55, 90, 104, 109, 118, 162, 182, 184, 193, 205	7	9, 13	Anthelmintic	Narcotic, scorpion sting, antiseptic, rheumatism.
41	*Ocimum sanctum* Linn Tulasi/Basil Lamiaceae	Essential oil— eugenol, carvacrol	–	4, 18, 25, 37, 45, 83, 100, 126, 146, 169, 172, 181, 184	5, 115, 166, 179, 186, 229, 232	10, 59, 99, 186, 189, 204, 208, 211, 229	–	Antiseptic, used in malarial fever, leaves snuff in Ozaena. In snake bite and scorpion sting.
42	*Parthenium hysterophorus* L Gajar gavat Asteraceae	Parthenin, partheniol, hymenin	–	4, 79, 98, 145, 210	128, 213	59, 149, 204	–	Tonic, febrifuge and emmenagogue.
43	*Peganum harmala* Linn Harmala Rutaceae	Alkaloid— harmine, peganine, harmaline, harmalol	150	31, 36, 37, 38, 39, 91, 92, 121, 182	103	103	Taenicidal	Used as narcotic and in rheumatism.
44	*Phyllanthus niruri* Linn Bhui-awala Euphorbiaceae	Bitter substances— phyllanthin.	32, 36, 37, 112	98	74, 221	74, 221	Molluscicidal	Toxic to fish and frog, fresh roots on jaundice and women menstrual, anti-tumour, antiviral.
45	*Polygonum hydropiper* Linn Pucker-mul Polygonaceae	Essential oil, glycoside.	36, 37, 112	36, 38, 93, 117, 119, 161, 223	–	–	–	Uterine disorders.
46	*Pongamia glabra* Vent. Karanj Leguminosae	Bitter fatty oil, essential oil karanjin.	32, 36, 37, 38, 83, 91, 112, 150	15, 27, 68, 79, 80, 95, 116, 122, 162, 181, 193, 199, 209, 214, 225	195	99, 195, 226	Wormicidal	Oil in piles, scabies, rheumatism, diabetes, skin diseases, antiseptic.
47	*Randia dumetorum* Lam Gelphal Rubiaceae	Saponin, essential oil, acid, resin	36, 37, 83, 91, 112, 148, 150	2, 37, 38, 91	–	–	Anthelmintic	Antiseptic, abortifacient, rheumatism, children teeting.

(Contd.)

TABLE 5.1 (*CONTINUED*)

No.	Botanical and vernacular name with family	Chemical ingredients	Biocidal activity reported					Medicinal uses
			Piscicidal	* Insecticidal	Anti-bacterial	Anti-fungal	Others	
48	*Ricinus communis* Linn Erand/Erandi Euphorbiaceae	Alkaloid—ricinine, toxalbumin, ricin.	36, 37, 38, 91, 112	25, 28, 38, 45, 55, 95, 101, 120, 122, 146, 164, 182, 191, 199, 214, 225, 239	142	–	Nematicidal	Scorpion sting, promotes lactation and menstrual discharge.
49	*Sapindus trifoliatus* Linn Ritha Sapindaceae	Saponin	2, 36, 37, 38, 91, 111, 112, 150	66, 102	–	33	–	Epilepsy, asthma, hysteria, hemicrania, hair washing, anti-tubercular.
50	*Sphaeranthus indicus* Linn Gorakhmundi Asteraceae	Alkaloid—sphaeranthine, essential oil and terpenoids.	36, 37, 39, 91, 136, 150	55, 76, 77, 185, 225	108, 113, 165	58, 99, 108,165, 168	Anthelmintic, crabicidal	Tonic, urethral discharge, germicidal, jaundice, piles, anti-tubercular, haemolytic, wound healing, anti-inflammatory.
51	*Spilanthes acmella* Murr. Akarkarha Asteraceae	Spilanthol, sterol	37, 112	155	–	–	–	Produce salivation in dry mouth, local anaesthetic action.
52	*Tagetes minuata* Linn Marigold /Zendu Asteraceae	Salicylic acid, bitter substances and essential oil—sesquiterpenes, volatile oil—tagetone.	–	21, 55, 96, 130, 138, 171, 184, 230, 233	229	138, 208, 226, 229	Nematicidal	Used in amenorrhoea, diaphoretic, wound and injuries treatment.
53	*Tephrosia purpurea* (Linn) Pers. Unhaallee Leguminosae	Rotenone, glucoside, rutin, tephrosin.	32, 36, 37, 38, 39, 83, 112, 150	145, 123, 240	53, 113	–	–	Anti-inflammatory and disorders of liver, kidney and spleen. Anti-ulcer, blood purifier, sex stimulant, diarrhoea, in hydrocoel, jaundice, diabetes and in veneral diseases of women.

No.	Plant name / Family	Chemical constituents						Uses
54	*Thevetia neriifolia* Juss Yellow kanher Apocynaceae	Glucoside—thevetin, thevetoxin.	36, 38, 112, 150, 177	72	–	33	Molluscicidal	Acronarcotic poison, cardiotonic, used in oedema, milky juice highly poisonous used in suicide, abortifacient.
55	*Trigonella foenumgraecum* Linn. Methi/Fenugreek Leguminosae	Alkaloid—trigonelline and choline, essential oil, saponin.	–	36, 38, 83	11, 26	33, 154	–	Tonic, cooling drink to small pox patients, rheumatism, dyspepsia.
56	*Tridax procumbens* Linn Ekdandi Asteraceae	Essential oil	–	145, 143, 184, 225	54,178, 215	–	–	Used to avoid infections.
57	*Vinca rosea* Linn Sadaphuli Apocynaceae	Alkaloid, essential oil	–	60, 61, 127, 146	37	33, 166, 208	–	Used in diabetes, wasp sting, heartpoison.
58	*Vitex nigundo* Linn Nirgudi Verbenaceae	Alkaloid—nishidine, essential oil, glucoside.	–	15, 28, 36, 38, 45, 55, 65, 83, 90, 100, 122, 126, 141, 146, 163, 173, 176, 203, 209, 210, 216, 218, 225	153	226	Wormicidal, anthelmintic	In malarial fever, rheumatism, germicidal, anti-parasitic, nerve tonic, antiseptic, on swelling.
59	*Withania somnifera* Dunal Ashwagandha Solanaceae	Alkaloid	–	12, 15, 60, 61, 65, 71, 203	68, 137	–	–	Narcotic, rheumatism, ulcer, local infections, in eye diseases, antiseptic, anti-tubercular, wound healing, anti-inflammatory
60	*Zingiber officinale* Rosc Adrak Zingiberaceae	Essential oil—gingerol, Zingiberine.	–	5, 17, 87, 100, 172, 173	232	59, 94, 154, 190, 200, 201	–	Antiseptic, tonic and stimulating remedy. Used in paralysis and rheumatism

*Insecticidal activity—the plant may act either in one or more way especially insecticidal, larvicidal, antifeedant, ovipositional deterrence, ovicidal, repellent or growth disruptor, etc.

their medicinal uses along with botanical and vernacular names with family, chemical ingredients, etc. are listed in Table 5.1. From this table, it is seen that all 60 plants are insecticidal, 29 plants are piscicidal, and 38 are antibacterial whereas 37 plants are antifungal. All these plants have folkoric reputation, used by earlier researchers. The literature survey of these plants as biocidal agents has been taken from various sources like

   i.   information from the herbalist,

   ii.   folkoric reputation,

   iii.   cross references and

   iv.   internet.

REFERENCES

Abdel-Aziz, Shadia, Omer, E.A. and Aziz, A. (1995). Bio-evaluation of *Dodonaea viscosa* L. Jacq extracts on the cotton leaf worm, *Spodoptera litteralis* (Biosd) as indicated by life table parameters. *Annals of Agricul.Sci. Cairo*. 40(2): 891–900.

Agarwal, V.S. and Ghosh, B. (1984). *Drug plants of India (Root Drugs)*. Kalyani Publishers, New Delhi.

Ahmed, S. and Graigne, M. (1985). The use of indigenous plant resources in rural development: Potential of neem tree. *Int. Development Technology*. 3: 123–130.

Ahmed, S.K. and Bhattacharya, A.K. (1991). Growth inhibitory effect of some plants for *Spirosoma oblique* Walker. *Indian Journal of Entomology*. 53(3): 453–474.

Ahmed, F.B.H., Mackeen, M.M., Ali, A.M., Mushiram, S.R. and Yaacob, M.M. (1995). Repellency of essential oils against the domiciliary Cockroach, *Periplaneta americana*. *Insect Science and its Application*. 16 (3/4): 391–393.

Ahsan, M. and Islam, S.N. (1996). Garlic: A broad spectrum antibacterial agent effective against common pathogenic bacteria. *Fitoterapia*. 67(4): 374–376.

Akinpelu, D.A. and Obuotor, E.M. (2000). Antibacterial activity of *Nicotiana tabacum* leaves. *Fitoterapia*. 71(2): 199–200.

Al-Lawati, H.T., Azam, K.M. and Deadman, M.L. (2002). Insecticidal and repellent properties of subtropical plant extracts against pulse beetle, *Callosobruchus chinensis*. *Sultan-Qaboos-University Journal for Scientific Research Agricultural Sciences*. 7(1): 37–45.

Amadioha, A.C. and Obilor, P.N. (2002). Fungitoxic effects of *Nicotiana tabacum* and *Carica papaya* on *Rhizopus oryzae* causing yam tuber rot in storage. *Journal of Sustainable Agriculture and Environment*. 4(1): 108–113.

Amadioha, A.C. (2004). Control of black rot of potato caused by *Rhizoctonia bataticola* using some plant leaf extracts. *Archives of Phytopathology and Plant Protection*. 37(2): 111–117.

Amalraj, A., Balasubramanian, A., Edwin, E. and Sheja, E. (2005). Antimicrobial activity of fenugreek seeds and leaves. *Indian Journal of Natural Products*. 21(2): 35–36.

Anandkumar, Dwivedi, N.S. and Dwivedi, S.C. (2004). Phagodeterrent activity of plant extracts against *Corcyra cephalonica* Stainton. (Lepidoptera: Pyralidae). *Pestology*. 28(6): 64–66.

Anonymous. (1975). Natural fungicides from tobacco suggest new approach for plant disease control. *World Crops*. Sep./Oct.: 237.

Anuradha, V., Lakshmi, T., Sakthivadivel, M. and Daniel, T. (2000). Effect of certain plant extracts against the fourth instar larvae of filarial vector *Culex quinquefasciatus*. *J. Ecobiology*. 12(2): 93–98.

Anuradha, V., Gowari, Neelima., Daniel, T. and Neelima, G. (2002). Effect of certain plant extracts on *Callosobruchus maculates* as pest on stored pulses. *J. Ecobiology*. 14(1): 67–71.

Archana Tiwari, Kumar, M.L. and Saxena, R.C. (1995). Effect of *Nicotiana tabacum* on *Trilobium castaneum*. *International Journal of Pharmacognosy*. 33(4): 348–350.

Areekul, S., Sinchaisri, P. and Tigvatan Anon, S. (1987). Effects of Thai plant extracts on the oriental fruit fly. *Kasetsart Journal, Natural-Sciences*. 21(4): 395–407.

Ashalata D' Rozario., Subir Bera and Dipak Mukherji. (1999). *A Handbook of Ethnobotany*. Kalyani Publisher, Ludhiana.

Asolkar, L.V., Kakkar, K.K. and Chakre, O.J. (2000). *Glossary of Indian Medicinal Plants with Active Principles*. Part-I (A–K). CSIR.

Atwal, A.S. and Pajni, H.R. (1964). Preliminary studies on the insecticidal properties of drupes of *Melia azadarach* against caterpillars of *Pieris brussicae*. *Indian J. Ento.* 26: 221–227.

Basabose, K., Bagalwa, M. and Chifundera, K. (1997). Anophelinocidal activity of volatile oil from *Tagetes minata* L. (Araceae). *Tropiculture*. (Congo). 15(1): 8–9.

Bhasin, H.D. (1926). Rept. Operations Dept. Agr. Punjab. 1: 69.

Bhatnagar Thomas, P.L. and Pal, A.K. (1974). Studies on the insecticidal activity of garlic oil. 1. Differential toxicity of the oil to *Musca domestica nebulo* Fabr. and *Trogoderma granarium* Everts. *J. Food Sciences and Technology* 1: 110–113.

Bhatnagar, A., Bhadamia, N.S. and Jakhmola, S.S. (2001). Efficacy of vegetable oils against pulse beetle *Callosobruchus maculatus* in cowpea. *Indian J. Entomology*. 63(3): 237–239.

Bhattacharjee, S.K. (2004). *Hand Book of Medicinal Plants*. Pointer Publishers, Jaipur.

Bhatti, M.A., Khan, M.T.J., Ahmed, B., Jamshaid, M. and Ahmed, W. (1996). Antibacterial activity of *Trigonella foenum-graecum* seeds. *Fitoterapia*. 67(4): 372–374.

Bhonde, S.B., Deshpande, S.G. and Sharma, R.N. (2001). Toxicity of some plant products to selected insect pests and its enhancement by combinations. *Pesticide Research Journal*. 13(2): 228–234.

Buiyah MIM and Quiniones A.C. (1990). Use of leaves of lagundi (*Vitex nigundo* Linn) as corn seed protectant against the corn weevil *Sitophilus zeamais* Motsch. *Bangladesh Journal of Zoology*. 18(1): 127–129.

Bunker, G.K. and Bhargava, M.C. (2002). Preliminary studies on the ovicidal effect of some vegetable oils against *Corcyra cephalonica* Stainton. *Insect Environment*. 8(3): 102–103.

Buthan, D.K. (1971). Studies on the efficacity of neem seed kernel powder against grain pests. *Indi-an Agri. Res. Inst*. Division of Entomol.

Caius, J.F. (1986). *The Medicinal and Poisonous Legumes of India*. Scientific Publishers, Jodhpur, India.

Caius, J.F. (1998). *The Medicinal and Poisonous Plants of India*. Scientific Publishers, Jodhpur, India.

Careaga, M., Fernandez, E., Dorantes, L., Mota, L., Jaramillo, E.M. and Hernandez-San Chez, H. (2003). Antimicrobial activity of *Capsicum* extract against *Salmonella typhimurium* and *Pseudomonas aeruginosa* inoculated in raw beef meat. *International Jour. of Food Microbiology*. 83(3): 331–335.

Chary, M.P., Reddy, E.J.S. and Reddy, S.M. (1984). Screening of indigenous plants for their antifungal principles. *Pesticides*. P. 17–18.

Chattoraj, A.N. and Tiwari, S.C. (1965). A note on the insecticidal property of *Annona squamosa*. Nat. Acad. Science (India) 35, Sec. B. Part IV: 351–353.

Chopra, R.N., Badhwar, R.L. and Ghosh, S. (1949). *Poisonous Plants of India*. Vol-I. The Indian Council of Agriculture Research, Delhi.

Chopra, R.N., Nayar, S.L. and Chopra, I.C. (1956). *Glossary of Indian Medicinal Plants*. CSIR, New Delhi.

Chopra, R.N., Chopra, I.C., Handa, K.L. and Kapur, L.D. (1982). *Chopra's Indigenous Drugs of India*. Academic Publishers: Calcutta.

Chopra, R.N., Badhwar, R.L. and Nayar, S.L. (1986). *Insecticidal and Piscicidal Plants of India*. CSIR, New Delhi.

Choudhari, R.K., Veda, O.P. and Mandloi, K.C. (2001). Use of neem, *Azadirachta indica* (L.) and Garlic, *Allium sativum* (L) in management of boll-worms in cotton. Proceedings of the 88th section of the Indian Science Congress, Agricultural Sciences, New Delhi. p. 40–42.

Chowdhury, B.K., Jha, S., Bhattacharyya, P. and Mukherjee, J. (2001). Two new carbezole alkaloids from *Murraya koenigii*. *Indian Jour. of Chemistry*. 40 B(6): 490–494.

Cooke, T. (1903). *The Flora of the Presidency of Bombay*. Published under the authority of the Secretary of State for India in council. Vol. I.

Dabur, R., Singh, H., Chhillar, A.K., Ali, M. and Sharma, G.L. (2004). Antifungal Potential of Indian Medicinal Plants. *Fitoterapia*, 75(3–4): 389–91.

Dar, A. and Ahmad, S. (2003). Efficacy of plant extracts against *Alternaria brassicae* (Berk) Sacc. and *Altenaria brassiciola* (Schw) Wilts. The incitant of altenaria leaf blight of knol-khol (*Brassicae oleraceae* var. *gongylodes*). National Seminar on organic products and their future prospects, SKUAST (K), Srinagar. p. 105.

David, B.V., Sukumaran, D. and Kandasamy, C. (1988). The Indian Privet *Vitex nigundo* Linn. A plant possessing promising pesticidal activity. *Pesticides.* 22(2): 27–30.

Dayal, R., Tripathi, R.A. and Renu. (2003). Comparative efficacy of some botanicals on protectant against *Sitophilus oryzae* in rice and its palatability. *Annals of Plant Protection Sciences.* 11(1): 160–162.

De, M., De, A.K. and Banerjee, A.B. (1999). Antimicrobial screening of some Indian spices. *Phytotherapy Research.* 13(7): 616–618.

Deena, M.J. and Thoppil, J.E. (2000). Antibacterial activity of the essential oil of *Lantana camera. Fitoterapia.* 71(4): 453–455.

Deka, M.K., Karan Singh., Handique, R. and Singh, K. (2001). Efficacy of wild sage (*Lantana camera* L.) and Basak (*Adhatoda vesica*) against tea mosquito bug in the field. *Research on Crops.* 2(1): 66–70.

Deokar, A.B. (1998). 125 *Medicinal plants Grown at Rajegaon.* Herbal Academy, DS Manav Vikas Foundation, Pune.

Deota, P.T. and Upadhyay, P.R. (2005). Biological studies of azadirachtin and its derivatives against polyphagous pest, *Spodoptera litura. Natural Product Research.* V. 19(5): 529–539.

Deshpande, R.S. and Tipnis, H.P. (1977). *Pesticides.* 11: 11–12.

Deshpande, S., Shah, T., Shah, G. and Parmar, N.S. (2005). A preliminary study on antimicrobial activity of *Tephrosia purpurea. Indian Journal of Natural Products.* 21(2): 33–34.

Devi, M., Vimala and Suneeta, A. (1990). Antibiotic efficacy of herbal ointments containing *Acalypha indica, Tridax procumbens* and *Azadirachta indica. J. Re. Ed. Ind. Med.* 9: 39–41.

Devkumar, C. and Sukhadev. (1993). Neem Research and development Radhwa, N.S. and Parmar, B.S. (eds.). Society of Pesticide Science, New Delhi. pp. 63–96.

Dhaliwal, G.S. and Arora, R. (2003). *Principles of Insect Pest Management.* Kalyani Publisher. Ludhiana. pp. 151–155

Drury, Colonel Herber. (1985). *Useful Plants of India.* International Book Distribution. Dehradun.

Dubey, K.S., Ansari, A.H. and Hardaha, M. (2000). Antimicrobial activity of the extract of *Sphaeranthus indicus. Asian J. of Chemistry.* 12 (2): 577–578.

Dubey, A.K., Pandya, R.K., Bartaria, A.M. and Shrivastava, A.K. (2002). Efficacy of plant extracts against *Altenaria tagetica. J. of Mycology and Plant Pathology.* 32(2): pp. 278.

Dwivedi, S.C. and Rajesh Kumar. (1999). Screening of responses of adult khapra beetle, *Trogoderma granarium* L. on ten plants species for possible repellent action. *Uttar Pradesh Journal of Zoology.* 19(3): 197–200.

Dwivedi, S.C. and Anand Kumar. (1999). Ovicidal activity of six plant leaf extracts on eggs of *Corcyra cephalonica* Stainton (Lepidoptera: Pyrallidae). *Uttar Pradesh Journal of Zoology.* 19(3): 175–178.

El-Lakwah, F.A., Darwish, A.A. and Halowa, Z.A. (1996). Toxic effect of extracts and powders of some plants against the cowpea beetle (*Callosobruchus maculatus* F.). *Annals of Agricultural Sciences*, (Egypt). 34(4): 1849–1859.

El-Naggar, MEA-El; Abdel-Sattar-MM; Mosqllam-SS; Naggar-MEA-El; Sattar-MM-Abdel. (1989). Toxicity of Colocynithin and hydrated Colocynithin from alcoholic extract of *Citrullus colocynthis* pulp. *Journal of Egyptian Society of Parasitology.* 19(1): 179–185.

Elnima, E.I., Ahmed, S.A., Mekkawi, A.G. and Mossa, J.S. (1983). The Antimicrobial activity of garlic and onion extract. *Pharmazie.* 38(11): 747–8.

Fazal, H. (2003). *Nontoxic and eco-friendly repellents of plant origin.* National Symposium on Emerging Trends in Indian Medicinal Plants; Lucknow. 40(5–O/1)

Feinstein, L. (1952). *Insect, The Year Book of Agriculture.* U.S. Dept. Agri., Washington D.C. p. 222.

Frank Beye. (1978). Insecticides from the vegetable kingdom. In: *Plant Research and Development.* Institute of Pharmaceutical Biology, University of Freiburg. 7: 13–31.

Furmanowa, W., Gajdzis-Kuls, D., Starosciak, B. and Stefunska, J. (1980). Antibacterial activity of *Withania somnifera* (L.) Dun. Organ cultivated *in vitro*. *Herba Polonica*. 44(4): 265–269.

Gahukar, R.T. (2000). Bioefficacy of plant products used for Citrus pest Management in India. *Pestology*. XXIV (2): 27–32.

Gakuru, S., Foua Bi, K. (1996). Effect of plant extracts on Cowpea weevil (*Callosobruchus maculatus* Fab) and the rice weevil (*Sitophilous oryzae* L.) *Cahiers-Agricultures*. 5(1): 39–42.

Garg, S.C. and Jain, R.K. (2003). Antimicrobial activity of the essential oil of *Curcuma longa* L. *Indian Perfumer*. 47(2): 199–202.

Gebreyesus, T. and Chapya, A. (1983). *Natural Products for Innovative Pest Management*. Whitehead, D.L. and Bower, W.S. (eds.). Pergamon Press, Oxford. p. 237–241.

Ghatak, S.S. and Bhusan, T.K. (1995). Evaluation on the ovicidal activity of some indigenous plant extracts on Bihar hairy caterpillar, *Spilosoma oblique* (wk.) Arctiidae:Lepidoptera). *Environment and Ecology*. 13(2): 294–296.

Grewal, R.C. (2002). *Medicinal Plants*. Campus Books International, New Delhi.

Gritsunapan, W., Wuchi-Udomlert, M., Caichompoo, W. (2000). Antibacterial and antifungal activities of Curcuminoids from *Curcuma longa* grown in Thailand. *Phytomedicine*. 7 (Suppl. II) : 85.

Hameed, S.V.S.A., Ahamed, T.A.N. and Shah D.S.M. (2003). Larvicidal activity of crude leaf extract of *Sphaeranthus indicus* in cutworm, *Spodoptera litura* (Lepidoptera: Noctuidae). *Bionotes*. 5:2, 46–47.

Hammed, S.V.S.A. and Shah, D.S. (2003). Effect of aqueous extracts of *Sphaeranthus indicus* against *Culex fatigans* wied (Diptera: Culicidae). *Journal of Experimental Zoology*, (India). 6(2): 279–284.

Harish, Chander., Ahuja, D.K., Nagender, A., Berry, S.K. and Chadner, H. (1998). Efficacy of plant materials against *Tribolium castaneum* (Herbst) in milled rice under bagged condition. *Journal of Insect Science*. 11(2): 133–136.

Hernandez, T., Canales, M., Avila, J.G., Duran, A., Cabullero, J., Romo de vivar, A. and Lira, R. (2003). Ethnobotany and antibacterial activity of some plants used in traditional medicines of Zopotitian de las Salinas Puebla (Mexico) *J. Ethnopharmacol.* 88(2–3): 181–188.

Hiremath, I.G. and Ahn, Y.J. (1997). Parthenium as a source of pesticide. First International Conference on Parthenium management, Dharwad, India. p. 86–89.

Ignacimuthu, S. (2004). Green pesticides for insect pest management. *Current Science.* 86(8): 1059–1060.

Islam, B.N., Schmutterer, H. (eds.)., Ascher, K.R.S. (1987). Use of some extracts from Meliaceae and Annonaceae for control of rice hispa, *Disladispa armigera* and the pulse beetle, *Callosobruchus chinensis.* Natural pesticides from the neem tree and other tropical plants. Proceedings of the 3rd International Conference, Nairobi, Kenya. 217–242.

Jacobson, Martin and Crosby, D.G. (1971). *Naturally occurring Insecticides.* Marcel Dekker, Inc. New York. p. 177–239.

Jain, S.K. (1991). *Contribution to Indian Ethnobotany.* Scientific Publishers, Jodhpur.

Jaswanth, A., Ramanathan, P. and Ruckmani, K. (2002). Evaluation of mosquitocidal activity of *Annona squamosa* leaves against filarial vector mosquito *Culex quinquefasciatus* Say. *Indian Journal of Expt. Biology.* 40: 363–365.

Jaswanth, A., Ramanathan, M., Ravindra Babu, S., Manimaran, S. and Ruckman, K. (2002). Evaluation of insecticidal activity of *Annona squamosa* against the storage pest *Sitophilus oryzae. Indian Drugs.* 39(5): 297–298.

Javaid, I. and Poswal, M.A.T. (1995). Evaluation of certain spices for the control of *Callosobruchus maculatus* (F.) (Coleoptera: Bruchidae) in cow pea seeds. *African Entomology.* 3(1): 87–89.

Jaykumar, M., John William, S. and Ignacimuthu, S. (2005). Evaluation of some plant extracts on the oviposition deterrent and adult emergence activity of *Callosobruchus maculatus* Fab. (Bruchidae: Coleoptera). *Pestology,* Mumbai. 29(1): 37–41.

Jha, M.M., Sanjeev Kumar and Saleh-Hasan. (2004). Effects of botanicals on Maydis leaf blight of Maize *in vitro. Annals of Biology.* 20(2): 173–176

Joseph Kumari, T.S.J. and Skaria, B.P. (1999). Medicinal plants in pest control. *Indian Journal of Arecanut, Spices and Medicinal Plants.* 1(4): 123–124.

Joshi, S.G. (2003). *Medicinal Plants*. Oxford and IBH Publishing Co. Pvt. Ltd, New Delhi.

Ju-Yun wei, Zhao- Bo Guang, Cheng- Xiaofeng, Bi-Qingsi, Ju-YW, Zhao-BG, Cheng-XF, Bi-QS. (2000). Bioactivities of six desert plants extracts to *Heliothis armigera* Hubner. *Journal of Nanijing Forestry University*. 24(5): 81–83.

Kalita, J., Gogoi, P., Bhattacharyya, P.R. and Handique, R. (2002). A study on effect of traditional storage protectants on oviposition of *Callosobruchus chinensis* and their damage to stored pulses. *Journal of Applied Zoological Researches*. 13(2–3): 251–254.

Kapoor, A. (1997). Antifungal activities of fresh juice and aqueous extracts of turmeric (*Curcuma longa*) and ginger (*Zingiber officinale*). *J. of Phytological Research*. 10(1–2): 59–62.

Ketkar, C.M. and Schmutterer, R. (ed.), Ascher, K.R.S. (1987). Use of tree-derived non-edible oils as surface protectants for stored legumes against *Callosobruchus maculatus* and *Callosobruchus chinensis*. Natural pesticides from the neem tree (*Azadirachta indica* A. Juss) and other plants. Proceedings of the 3rd International Neem Conference, Nairobi, Kenya. p. 535–542.

Khan, M.A., Tiwari, S. and Joshi, B. (2001). Some herbs: Ovicides of Corcyra cephalonica Stainton. National Research Seminar on Herbal Conservation, Cultivation, Marketing and Utilization with special emphasis on Chhatisgarh, The Herbal State, Raipur, Chhatisgarh. p. 111.

Khan, S., Jabbar, A., Hasan, C.M., Rashid, M.A. (2001). Antibacterial acitivity of *Barringtonia racemosa*. *Fitoterapia*. 72(2): 162–164.

Kirti-Sharma., Saxena, D.B., Gupta, A.K. and Sharma, K. (2003). Bioefficacy of *Tinospora cordifolia* and *Phyllanthus niruri* plants against *Spodoptera litura* (F.) and *Dysdercus koenigii* (Fabricus). *Pesticide Research Journal*. 15(20): 138–142.

Kishore, G.K., Pande, S. and Rao, J.N. (2001). Control of late leaf spot of groundnut (*Arachis hypogaea*) by extracts from non-host plant species. *Plant Pathology Journal*. 17(5): 264–270.

Kokate, S.D. and Chintalwar, G.J. (2003). Insect repellent activity of certain plant extracts against pulse beetle *Callosobruchus chinensis* L. *National-Academy-Science-Letters*. 26(1/2): 44–46.

Kumar, A. and Datta, G.P. (1987). Indigenous plant oils as larvicidal agents against *Anopheles stephenst* mosquitoes. *Current Science*. 56(18): 959–960.

Kumar, B.H. and Thakur, S.S. (1988). Certain non-edible seed oils as feeding deterrents against *Spodoptera litura* Fab. *Journal of the Oil Technologist Association of India*. 20(3): 63–65.

Kumar, V., Shah, T., Parmar, N.S., Shah, G.B. and Goyal, R.K. (2005). Antimicrobial activity of *Peganum harmala*. *Ind. J. Natural Products*. 21(2): 24–26.

Kumari, P. and Kumar, D. (1998). Effect of mixture of tobacco leaf and neem seed powder on *Callosobruchus chinensis* L. infesting pulse grains. *J. of Ecotoxicology and Environmental Monitoring*. 8(4): 229–232.

Kumbhar, P.P., Salunkhe, D.H., Borse, M.B., Hiwale, M.S., Nikam, L.B., Bendre, R.S., Kulkarni, M.V. and Dewang, P.M. (2000). Pesticidal potency of some common plant extracts. *Pestology*. XXIV (6): p. 51–53.

Lavie, D., Jain, J.K. and Shapan Garieleith, S.R. (1967). *Chem. Commun*. p. 910.

Lirio, L.G., Hermano, M.L. and Fontanilla, M.Q. (1998). Antibacterial activity of medicinal plants from Philippines. *Pharmaceutical Biology*. 36(5): 357–359.

Lulla, P.M., Deshpande, A.R., Musaddiq, M., Shahare, N.H. and Ukesh, C.S. (2005). Antimicrobial activity of medicinal plant *Sphaeranthus indicus*. International Conference on Modern Trends in Plant Sciences with Special Reference to the Role of Biodiversity in Conservation. Amravati, Maharashtra. p. 96.

Madanlar, N., Yoldas, Z., Durmusoglu, E. and Gul, A. (2002). Investigation on the natural pesticides against pests in vegetable green houses in Izmir (Turkey). *Turkya-Entomoloji Dergisi*. 26(3): 181–195.

Madhiazhagan, K., Ramadoss, N. and Anuradha, R. (2002). Effect of botanicals on bacterial blight of rice. *Journal of Mycology and Plant Pathology*. 32(1): 68–69.

Mahajan, R.T., Choube, S.M. and Jawale, S.M. (1989). Piscicidal activity of some indigenous plants. *Perspective in Aquatic Biology*. p. 369–375.

Mahajan, R.T. (1994). Review on fish toxicants of plant origin. *Indian Journal of Environment and Toxicology*. 4: 7–17.

Mahajan, R.T., Chaudhari. G.S. and Chopda, M.Z. (1999). Screening of some indigenous plants for their possible Antibacterial activity. *Environmental Bulletin*. 15: 61–62.

Maheshwari, V.L., Kotkar, H.M., Mendki, P.S., Sangeetha, Sadan., Jha, S.R. and Upasani, S.M. (2001). Antimicrobial and Pesticidal activity of partially purified flavonoids of *Annona squamosa*. *Pest Management Sci*. 58: 33–37.

Majumdar, D.K., Singh, S. and Malhotra, M. (2005). Antibacterial activity of *Ocimum sanctum* L. fixed oil. *Indian Journal of Experimental Biology*. 43: 835–837.

Mallapur, C.P., Hulihalli, U.K. and Kubsad, V.S. (2001). Safflower aphid management through botanical pesticides. *Karnataka Journal of Agricultural Sciences*. 14(2): 321–325.

Manson, D.J. (1939). *Malaria Inst*. India. 2: 85.

Mareggiani, G. (2001). Management of insect pests with semiochemical substances originating from plants. *Manejo-Integrado-de-plagas*. 60: 22–30.

Mayabini, Jena and Jena, M. (2000). Efficacy of the plant *Polygonum hydropiper* against rice brown plant hopper *Nilaparvata lugens* Stal. *Current Science*. 78(8): 953–954.

Meena, B.L. and Bhargava, M.C. (2003). Effect of plant products on reproductive potential of *Corcyra cephalonica* Stainton (Lepidoptera: Pyralidae). *Annals of Plant Protection Sciences*. 11(2): 196–200.

Meera, Srivastava, Mann, A.K. and Shrivatava, M. (2002). An evaluation of efficacy of extracts of plant *Peganum harmala* against pulse beetle *Callosobruchus chinensis*. *Indian Journal of Entomology*. 64(2): 138–147.

Meshram, P.B. (2000). Antifeedant and insecticidal activity of some medicinal plant extracts against *Dalbergia sissoo* defoliator *Plecoptera reflexa* Gue. (Lepidoptera: Noctuidae). *Indian forester*. 126(9): 961–965.

Minja, E.M., Silim, S.N. and Karuru, O.M. (2002). Efficacy of *Tephrosia vogelli* crude leaf extract on insect feeding on pigeon pea in Kenya. *International Chickpea and Pigeon pea*. *Newsletter*. 9: 49–51.

Mironow, V.S. (1940). The use of *Acorus calamus* for insecticidal and repellent preparations. *Med. Parasitol*. 9: 409–410.

Mishra, R.C., Mishra, D.B. and Gupta, P.R. (1984). Effects of mixing of *Mentha* powder with and giving oil in water emulsion dip to the Chickpea on pulse beetle *Callosobruchus chinensis* L. *Bull. Grain Tech*. 22(1): 19–23.

Misra, H.P. (2000). Effectiveness of indigenous plant products against the pulse beetle *Callosobruchus chinensis* on stored black gram. *Indian Journal of Entomology*. 62(2): 218–220.

Mohsen, Z.H., Abdul-Latif, M.J., Al-Chalabi, B.M. and Al-Naib, A. (1989). Insecticidal activity of *Vinca rosea* against *Culex quinquefasciatus*. *Journal of Biological Sciences Research*. 20(3): 437–446.

Montanaro, S., Bardon, A. and Catalan, C.A.N. (1996). Antibacterial activity of various Sesquiterene lactones. *Fitoterapia*. 67(2): 185–187.

Mulla, M.S. and Su, T. (1999). Activity and biological effects of neem products against arthropods of medical and veterinary importance. *J. Am. Mosq. Control Assoc*. 15(2): 133–52.

Munyulli-bin-Mushambanyi, T. (2003). The utilization of botanical dusts in the control of foodstuff storage insect pests in Kivu (Democratic Republic of Congo). *Cahiers Agricultures*. 12(1): 23–31.

Navi, S.S. and Singh, S.D. (2004). Effects of pounding and garlic extract on Sorghum grain mold and grain quality. *International-Sorghum and Millets-Newsletter*. 44: 122–124.

Nayak, S. and Singhai, A.K. (2003). Antimicrobial activity of the roots of *Cocculus hirsutus*. *Ancient Science of life*. 22(3): 101–106.

Neela Singh and Sushil Kumar. (1999). Screening of saponin glycosides, isolated from *Madhuca indica* seeds against forest pests. *Ann. Entomol*. 17(1): 45–47.

Ohsawa, K., Kato, S., Honda, H. and Yamamoto, J. (1990). Pesticidal active substances in tropical plants-insecticidal substances from the seeds of Annonaceae. *Journal of Agricultural Sciences, Tokyo-Nogyo-Daigaku*. 34(4): 253–258.

Osuala, F.O. and Okwuosu, V.H. (1993). Toxicity of *Azadirachta indica* to fresh water snails and fish with reference to the physiochemical factor effect on potency. *Appl. Parasitol*. 34(1): 63–8.

Oudhia Punkaj. (2003). Research Notes- Traditional medicinal knowledge about herbs and insects. Botanical.com.

Owais, M., Sharad, K.J., Faisal, S.M., Khan, M.A. and Saleemuddin, M. (2003). Antibacterial efficacy of *Withania somnifera* against experimental *Salmonella typhimurium* infection in balb/C mice. 2nd World Congress on "Biotechnological Developments of Herbal Medicines." NBRI, Lucknow, (20–22): 96.

Padma Vasudevan, Suman Kashyap and Satyavati Sharma. (1997). *Tagetes*: A multipurpose plant. *Bioresources Technology*. 62(1/2): 29–35.

Pal, R.K., Tripathi, R.A. and Prasad, R. (1996). Relative toxicity of certain plant extracts to Khapra beetle, *Trogoderma granarium*. *Annals of Plant Protection Sciences*. 4(1): 35–37.

Pandey, A.K. (2003). Composition and *in vitro* antifungal activity of the essential oil of menthol mint (*Mentha arvensis* L.) growing in Central India. *Indian Drugs*. 40(2): 126–128.

Paneru Ram, B. and Shivakoti Gopal, P. (2001). Use of Botanicals for the management of pulse beetle (*Callosobruchus maculatus* F.) in Lentil. Nepal *Agricultural Research Journal*. 4.

Parameswari, C.S. and Tulasi Latha, A. (2001). Antibacterial activity of *Ricinus communis* leaf extract. *Indian Drugs*. 38(11): 587–588.

Pathak, A.K. and Dixit, V.K. (1988). Insecticidal and insect repellent activity of essential oils of *Tridax procumbense* and *Cyathocline lyrata*. *Fitoterepia*. 59(3): 211–214.

Pathak, N., Yadav, B.P. and Vasudevan, P. (2000). Anti-termites characteristics of *Murraya* (Curry Patta) leaves. *Indian Jour. of Entomology*. 62(4): 439–441.

Patil, K.J., Chaudhari, V.A., Rane, A.E. and Nibalkar, S.A. (1993). Explorations of insecticidal properties in some weed plant materials against *Dactynotus carthami* H.R.L. and *Dysdercus koenigii* F. *Botanical pesticides in integrated pest management*. pp. 245–251.

Patil, R.S. and Goud, K.B. (2002). Preliminary studies on ovicidal activity of methanolic plant extracts against diamondback moth (DBM). *Plutella xylostella* (L.). *Insect Environment*. 8(2): 105–106.

Patil, V.J., Deshmukh, M.B. and Maner, M.I. (2002). Antimicrobial and antifeedant activity of the extract of the plant *Duranta repens*. *Biotechnology in Agriculture Industry and Environment*. Deshmukh, A.M. (ed.). Microbiologist Society, Karad. pp. 65–67.

Patil, D.A. (2003). *Flora of Dhule and Nandurbar Districts. (Maharashtra)*. Bishen Singh Mahendra Pal Singh, Dehradun.

Patni, C.S., Kolte, S.J. and Awasthi, R.P. (2005). Efficacy of botanicals against Alternaria blight (*Alternaria brassicae*) of mustard. *Indian Phytopath*. 58(4): 426–430.

Patole, S.S. and Mahajan, R.T. (2006). Phytotoxicity in fishes In: *Advances in Fish Research*. Vol. 4. Datta Munshi, J.D.S., (ed.). Narendra Publication, New Delhi. pp. 299–312.

Patro, B. and Sethi, P.M. (1997). Repellent effect of some plant extracts against the Pulse beetle (*Callosobruchus chinensis* Linn.) infesting green gram seeds. *J. of Appl. Zoological Researches*. 8(1): 20–27.

Pauli, F.F., Opazo, M.A.U. and Nobrega, L.H.P. (2002). The study of the effect of repellent plants to insect in the Corn seeds physiological quality stored in corn-cob through a statistic longitudinal analysis. *Revista-Brasileira-de-Produtos-Agroindustriasis*. 4(2): 167–174.

Perumal samy, R., Ignacimuthu, S. and Sen, A. (1998). Screening of 34 Indian Medicinal Plants for antibacterial properties. *J. Ethnopharmacol*. 62(2): 173–82.

Perumal, G., Subramanyam, C., Natrajan, D., Srinivasan, K., Mohanasundari, C. and Prabhakar, K. (2004). Antifungal activities of traditional medicinal plant extracts: A Preliminary survey. *J. Phytological Research*. 17(1): 81–83.

Pitasawat, B., Choochote, W., Kanjanapothi, D., Panthone, A., Jitakdi, A. and Chaithong, V. (1998). Screening for larvicidal activity of ten carminative plants. *Southeast Asian J. Trop Med Public health*. 29(3): 660–2.

Prabu Seenivasan, S., Jayakumar, M., Raja, N. and Ignacimuthu, S. (2004). Effect of bitter apple, *Citrullus colocynthis* (L.) Schard seed extracts against pulse beetle, *Callosobruchus macualtus* Fab (Coleoptera: Bruchidae). *Entomon*. 29(1): 81–4.

Pranata, R.I. (1985). Possibility of using turmeric (*Curcuma longa* L.) for controlling storage insects. *Biotrop. Newletter*. 45: 3.

Punam Kumari, Prem Kumari, Verma, M.K., Kumari P. and Kumari P. (1999). Effect of *Acorus calamus* as pulse grain protectant against *Callosobruchus chinensis*. *Journal of Applied Biology*. 9(2): 188–189.

Qureshi, S., Rai, M.K. and Agrawal, S.C. (1997). *In vitro* evaluation of inhibitory nature of extracts of 18 plants species of Chhindwara against 3 keratinophilic fungi. *Hindustan Antibios Bull.* 39(1–4): 56–60.

Rahman, G.K.M.M. and Motoyama, N. (2000). Repellent effect of garlic against stored product pests. *Journal of Pesticide Science.* 25(3): 247–252.

Rahman, Inee Gogoi, Dolui, A.K. and Ruma Handique. (2005). Toxicological study of plant extracts on termite and laboratory animals. *J. Environ. Biol.* 26(2): 239–241.

Rai, H.S., Verma, M.L. and Gupta, M.P. (1997). Efficacy of some indigenous plant products against the incidence of leaf roller and capsule borer, *Antigastra catalaunalis* Dup in Sesame. *Ann. Entomol.* 15(2): 27–31.

Raja, N., Albert, S. and Ignacimuthu, S. (2000). Effect of solvent residue of *Vitex nigundo* Linn and *Cassia fistula* Linn on Pulse beetle, *Callosobruchus maculates* Fab. and its larval parasitoid *Dinarmus vagabundus* (Timberlake). *Indian Jour. of Expt. Biology.* 38(3): 290–292.

Ram Singh, Basant Singh and R.A. Verma. (2001). Efficacy of different indigenous plant products as grain protectant against *Callosobruchus chinensis* Linn on pea. *Indian J. Entomology.* 63(2): 179–181.

Ram, A.J., Bhakshu, L.M. and Raju, R.R.V. (2004). *In vitro* antimicrobial activity of certain medicinal plants from Eastern Ghats, India used for skin diseases. *Journal of Ethnopharmacology.* 90(2–3): 353–357.

Ramabhadra Raju, M., Krishna Murthy, K.V.M. and Subhu Reddy, C. (2003). Efficacy of plant extracts in the Management of Collar Rot of Groundnut (*Arachis hypogaea* Linn) caused by *Aspergillus nigar* V. Teighem. *The Andhra Agric. J.* 50(3 and 4): 287–290.

Rani, P. and Khullar, N. (2004). Antimicrobial evaluation of some medicinal plants for their antienteric potential against multi-drug resistance *Salmonella typhi. Phytother Res.* 18(8): 670–673.

Rao, B.G.V.N. and Joseph, P.L. (1971). Activity of some essential oils towards Phytopathogenic fungi. *Riechest Aromas Koer.* 21: 405–410.

Rao, M.S., Pratibha, G. and Korwar, G.R. (2000). Evaluation of aromatic oils against *Helicoverpa armigera. Annals of Plant Protection Sciences.* 8(2): 236–238.

Redknap, R.S. (1981). Field trials using locally prepared insecticides. (Neem-*Azadirachta indica*). Bunjul Christian Council of the Gambia.

Rodrigue, S., Pelicana, A., Heck, G. and Delfino, S. (1999). Evaluation of the effect of natural extracts of *Tagetes* spp. as bioinsecticies on adults of *Sitophilus oryzae* L. (Coleoptera: Curculionidae). *IDESIA.* 17: 79–89.

Rohan Rajapakse, Rajapakse, H.L. de, Z. and Disna Ratnasekera. (2002). Effect of botanicals on oviposition, hatchability and mortality of *Callosobruchus maculates* L. (Coleoptera: Bruchidae). *Entomon.* 27(1): 93–98.

Sahayaraj, K. and Saker, R. (1996). Efficacy of plant extracts against tobacco caterpillar larvae in groundnut. *International Arachis Newsletter.* 16: 38.

Saikia, D., Santha kumar, T.R., Kahol, A.P. and Khanuja, S.P.S. (1999). Comparative bioevaluation of essential oils of three species of *Cymbopogan* for their antimicrobial activities. *Jour. of Medi. and Aromatic Plant Sciences.* 21(1): 24.

Saleh, M., Kamel, A., Li, X., Swaray, J. (1999). Antibacterial triterpenoids isolated from *Lantana camera*. *Pharmaceutical Biology.* 37(1): 63–66.

Saminathan, V.R. and Jayaraj, S. (2001). Evaluation of Botanical pesticides against the mealybug, *Ferrisia virgata* Cockrell (Homoptera: Pseudococcidae) on cotton. *Madras-Agricultural Journal.* 88(7–9): 535–537.

Sambasivam, S., Karpagam, G., Chandran, R. and Khan, S.A. (2003). Toxicity of leaf extracts of yellow oleader *Thevetia nerifolia* on Tilapia. *J. Environ. Biol.* 24(2): 201–204.

Samy, R.P. and Ignacimuthu, S. (2000). Antibacterial action of some folklore medicinal plants used by tribals of Western Ghats of India. *J. Etanopharmacol.* 69(1): 63–71.

Saroja, S., Usha, K., Shoba, P. and Meenakumari, V. (1997). Biochemcial profile of selected patients with tuberculosis and bacterial activity of certain indigenous plants on Tubercle bacilli. *Indian Jour. of Nutrition and Dietetics.* 34(8): 193–198.

Satyavati, G.V., Raina, M.K. and Sharma, M. (1976). *Medicinal Plants of India,* Vol. I (A–G) ICMR, New Delhi.

Satyavati, G.V., Gupta, A.K. and Neeraj Tandon. (1987). *Medicinal Plants of India,* Vol. II (H–P) ICMR, New Delhi.

Savant, S.Y. (1974). *Maharashtratil Divya Vanaspati*. Continental Prakashan, Pune.

Saxena, R.C., Dixit, O.P. and Harsar, V. (1992). Insecticidal action *Lantana camera* against *Callosobruchus chinesis* (Coleoptera: Bruchidae). *J. Stored Prod. Res.* 28: 279–287.

Saxena, R.C. and Shrivastava, R.K. (1993). *Insect Pest Management through Natural Products*. Ashish Publishing House, New Delhi. pp. 13–30.

Saxena, R.C. and Archana Tiwari. (1993). Repellent and feeding deterrent activity of *Sphaeranthus indicus* against *Tribolium castaneum* (Herbst). *Bioscience Research Bulletin*. 9(1–2): 57–59.

Sengupta, S., Ghosh, S.N., Ghosh, S.B. and Das, A.K. (2004). Bioefficacy of some plant extracts against microorganisms. *Journal of Mycopathological Research*. 42(1): 31–34.

Senguttum, K., Kuttalam, S. and Srinivan, T. (2005). Bioefficacy of *Melia dubiacav* and neem products against major insect pests of tomato. *Pestology*. 29(1): 47–50.

Shamim, S., Ahmed, S.W. and Azhar, I. (2004). Antifungal activity of *Allium, Aloe* and *Solanum* species. *Pharmaceutical Biology*, 42(7): 491–498.

Shanmugam. (2004). *Ann. Plant Protect.* 12(1): 83–86.

Sharma, P., Singh, S.D., Rawal, P. and Lodha, P.C. (2002). Antifungal activities of plant extracts and oils against seed borne pathogens of Pea. *Jour. of Mycology and Pathology*. 32(1): 151.

Sharma, A.K., Baruah, K. and Bhardwaj, A.C. (2005). Mosquito larvicidal characteristics of certain phytoextracts. *Journal of Experimental Zoology*, India. 8(1): 109–112.

Shrivastava, S.D. (1986). Limnoids from the seeds of *Melia azedarach. J. Nat. Prod.* 49(1): 56–61.

Shrivastava, R.N., Rai, H.S., Verma, M.L. and Gupta, M.P. (1977). Efficacy of some indigenous plant products against the incidence of leaf roller and capsule borer, *Antigastra catalaunalis* Dup in Sesame. *Ann. Entomol.* 15(3): 27–31.

Shukla, A.C., Shahi, S.K. and Anupam Dikshit. (2002). *Eucalyptus pauciflora*—A potential source of sustainable eco-friendly storage pesticides. *Biotechnology of Microbe and Sustainable Utilization*. pp. 93–107.

Simin, K., Ali, Z., Khalid-uz-zaman, S.M. and Ahmed, V.U. (2002). Structure and biological activity of a new rotenoids from *Pongamia pinnata*. *Natural Product Letters*. 16(5): 351–357.

Singh, U.P., Singh, H.B. and Singh, R.B. (1980). The Fungicidal effect of neem (*rachta indica*) extracts on some soil-borne pathogens of grain (*Cicer arietinum*) *Mycologia*. 72: 1077–1093.

Singh, R. and Reena. (2003). Insecticidal properties of garlic, *Allium sativum*: A review. *Journal of Medicinal and Aromatic Plant Sciences*. 25(4): 1024–1038.

Singh, S.C. (2003). Effect of Neem leaf powder on infestation of the pulse beetle *Callosobruchus chinensis* in stored Khesari. *Indian J. Entomology*. 65(2): 188–192.

Singh, P.K. (2003). Effect of some oils against pulse beetle *Callosobruchus chinensis* in infesting pigeon pea. *Indian J. Entomology*. 65(1): 55–58.

Singh, G., Maurya, S., Catalan, C. and de Lampasona, M.P. (2005). Studies on essential oils, part 42: chemical, antifungal, antioxidant and sprout suppressant studies on ginger essential oil and its oleoresin. *Flavour and Fragrance Journal*. 20(1): 1–6.

Singh, I. and Ved Pal Singh. (2005). Effect of plant extracts on mycelial growth and aflatoxin produced by *Aspergillus flavus*. *Indian Journal of Microbiology*. 45(2): 139–142.

Singla, R. and Khan, M.N. (2004). Antimicrobial activity of *Lantena camera* root and stem extracts. Biosciences, *Biotechnology Research Asia*. 2(2): 123–126.

Singhvi, P.M., Yogita Lohar, Mamia Panwar, Tarun Sablok, Lohra, Y., Panwar, M. and Sablok, T. (2002). Repellent activity of plant extracts against lesser grain borer, *Rhyzopertha dominica* Fabr. *Journal of Applied Zoological Researches*. 13(2–3): 260–261.

Sinha, A.K., Verma, K.P., Agrawal, K.C. and Thakur, M.P. (2002). Antifungal activities of different plant extracts against *Collectotrichum capsici*. *J. of Mycology and Plant Pathology*. 32(2): 268.

Smith, A.E. and Secoy, D.M. (1981). Plants used for agricultural pest control in Western Europe before 1850. *Chemistry and Industry*. 1: 12–17.

Spickett, R.G.W. (1955). The chemistry of some lesser known insecticides of plant origin. *Colonial Plant and Animal Products*. 5: 288–303.

Spies, J.R. (1933). *J. Econ. Entomol.* 26: 285.

Srivastava, A. and Srivastava, M. (1998). Fungi toxic effect of some medicinal plants (on some fruit pathogens). *Philippine Jour. of Science.* 127(3): 181–187.

Sundararajan, G. and Kumuthakalavalli, R. (2000). Some indigenous insecticidal plants of Dindigul district, Tamil Nadu. *Journal of Eco-biology.* 12(2): 111–114.

Sundarajan. G. (2002). Control of Caterpillar *Helicoverpa armigera* using botanicals. *Journal of Ecotoxicology and Environmental Monitoring.* 12(4): 305–308.

Suriyavathanna, M. and M.S. Gomathi. (2005). Screening for antifungal activity of selected medicinal plants. *J. Ecobiol.* 17(5): 445–449.

Suryadevara P. and Khanam, S. (2002). Screening of plant extracts for larvicidal activity against *Culex quinquefasciatus. Jour. of Nat. Remedies.* 2(2): 186–188.

Swapna Sree, D. and Sreeramulu, A. (2002). Inhibitory effect of plant extracts on the management of bacterial leaf spot pathogen (*Xanthomonas axonopodis* pv. *vesicatoria*) infecting *Capsicum frutescnes* L. *Jour. of Nat. Remedies.* 2(2): 191–194.

Swaroop Singh and Gireesh Sharma. (2003). Efficacy of different oils as grain protectants against *Callosobruchus chinensis* in green gram and their effect on seed germination. *Indian J. Entomology.* 65(2): 500–505.

Taddei, A. and Rosas Romero, A.J. (2000). Bioactivity studies of extracts from *Tridax procumbens.* *Phytomedicines.* 7(3): 235–238.

Tandon Vishal, R. (2005). Medicinal uses and biological activities of *Vitex nigundo. Natural Product Radience.* 4(3): 162–165.

Tanzubil, P.B. (1987). The use of neem products in controlling the Cowpea weevil *Callosobruchus maculates* In: Schmutterer, H. and Ascher, K.R.S. (eds). Natural pesticides from the neem tree and other tropical plants, Proceedings of the 3rd International Neem Conference. Nairobi, Kenya. 517–523.

Tewary, D.K., Santosh, V. and Vasudevan, P. (2003). Bioefficacy of *Vitex nigundo* L. (Verbenaceae) plant products as protectant against *Sitophilus oryzae* (L.) (Coleoptera: Curculionidae). *Shashpa.* 10(2): 161–166.

Thadepalli, H., Bansal, M.B. and Vadad, A.F. (2005). Antibacterial activities of plant extracts. *Recent Progress in Medicinal Plants.* Vol. 9. *Plant Bioactives in Traditional medicines.* Hajumdar, D. K. *et al.* (eds.) Studium Press, LLC, USA. pp. 4–5.

Thomas, E., Shanmugam, J. and Rafi, M.M. (1996). Antibacterial activity of plants belonging to Zingiberaceae family. *Biomedicine*. 16(2): 15–20.

Thomas, E., Shanmugham, J. and Rafi, M.M. (1998). *In vitro* antimicrobial activity of certain medicinal plants. *Biomedicine*. 19(3): 185–190.

Tiwari, S. and Singh, A. (2004). Toxic and sub-lethal effects of oleandrin on biochemical parameters of fresh water air breathing murrel, *Channa punctatus* (Bloch). *Indian J. Expt. Biology*. 42(4): 413–418.

Tripathi, A.K., Jain, D.C. and Singh, S.C. (1999). Persistency of bioactive fractions of Indian plants, *Polygonum hydropiper* as an insect feeding deterrent. *Phytotherapy Research*. 13(3): 239–241.

Tripathi, A.K., Prajapati, V., Agrawal, K.K. and Sushil Kumar. (2000). Effect of Volatile oil constituents of *Mentha* species against the stored grain pests, *Callosobruchus maculatus* and *Tribolium castaneum*. *Journal of Medicinal and Aromatic Plant Sciences*. 22(1B): 549–556.

Tripathy, M.K., Sahoo, P., Das, B.C. and Mohanty, S. (2001). Efficacy of botanical oils, plant powders and extracts against *Callosobruchus chinensis* Linn attacking black gram (Cv. T9). *Legume-Research*. 24(2): 82–86.

Usha Dev. C., Devakumar, P.C., Agrawal, Jitendra Mohan, Joshi, K.D. and Indra Rani. (2002). Antifungal effect of Vitex *nigundo, Calatropis* spp. and other plant extracts against seed-borne fungi. *Pesticide Research Journal*. 14(2): 229–233.

Verma, S.P., Kanaiyia, R.S. and Upadhyay, R.R. (1998). Antifungal activity of some Euphorbiaceous plants. *Indian Journal of Plant Pathology*. 16 (1 and 2): 62–63.

Verma, S.K., Verma, R.K. and Saxena, R.D. (2004). Bioefficacy of herbal extracts against building termites, *Microceratermes beesoni*. *Pestology*. 28(6): 19–22.

Vidya, S.S., Dharmagadd, Mamata Tandonb and Padma Vasudevan. (2005). Biocidal activity of the essential oils of *Lantena camera, Ocimum sanctum* and *Tagetes patula*. *Journal of Scientific and Industrial Research*. 64: 53–56.

Vyas, B.N., Ganesan, S., Raman, K., Godrej, N.B., Mistry, K.B. (1999). Effect of three plant extracts and achook: a commercial neem formulation on growth and development of three noctuid pests. *Azadirachta-indica. A-Juss*. Singh, R.P. and Saxena, R.C. (eds.). Science Publishers, Inc., Enfield, USA. pp. 103–109.

Wada, K. and Munakata, K. (1968). Naturally occuring insect control chemicals. *J. Agr. Food Chem.* 16: 471.

Wannissorn, B., Jarikasem, S., Siriwangchai, T. and Thubthimthed, S. (2005). Antibacterial properties of essential oils from Thai medicinal plants. *Fitoterapia.* 76(2): 233–236.

Wells, C., Bertsch, W. and Perish, M. (1992). Isolation of volatiles with insecticidal properties from the genus *Tagetes* (Marigold). *Chromatographia*, 34(5–8): 241–248.

Worsley, R.R. (1934). *Ann. Appl. Biol.* 21: 649.

Yadav, R., Srivastava, V.K., Chandra, R. and Singh, A. (2002). Larvicidal activity of latex and stem bark of *Euphorbia tirucalli* plants on the mosquito *Culex quinquefasciatus. J. Comman Dis.* 34(4): 264–269.

Yadava, R.L. (1971). Use of essential oil of *Acorus calamus* L. as an insecticide against the pulse beetle *Bruchids chinensis L. Zangew. Ent.* 68: 289–294.

Yamasaki, R.B., Ritland, T.G., Barnby, M.A. and Klockre, J.A. (1988). Isolation and purification of salanin from need seeds and its quantification in neem and chinaberry seeds and leaves. *Journal of Chromatography.* (U.S.A.). 447(1): 277–283.

Yanar, Y., Kadioglu, I., Katluk, N.D., Casmeli, I. and Hangun, A. (2001). Fungicidal activity of some plant extracts against different plants pathogenic fungi. *Turkiye-Herboloji-Dergisi.* 4(1): 58–63.

Yogita Lohra., Singhvi, P.M. and Lohra, Y. (2000). Effectiveness of biopesticides on the development of *Tribolium castaneum* Herbst. infecting stored sorghum. *Journal of Applied Zoological Researches.* 11(2–3): 128–131.

Zhang-YeGuang, Xu-HanHong, Huang-Ji Guang, Chiu-ShinFoon, Zhang-YG, Xu-HH, Huang, J.G. and Chiu, S.F. (2000). Antifeedant activity of *Tephrosia vogelli* (Hook) against species of Lepidoptera. *Journal of South China Agricultural University.* 24(4): 26–29.

www.ingramcontent.com/pod-product-compliance
Lightning Source LLC
Chambersburg PA
CBHW060113120726
48003CB00009B/2622